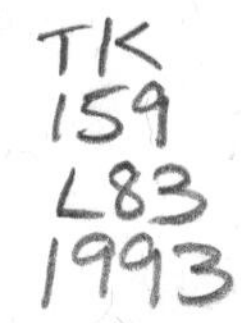

Lucky Strikes . . . Again

Lucky Strikes . . . Again

Robert W. Lucky

The Institute of Electrical and Electronics Engineers, Inc., New York

IEEE PRESS
445 Hoes Lane, PO Box 1331
Piscataway, New Jersey 08855-1331

This book may be purchased at a discount from the publisher when ordered in bulk quantities. For more information contact:

IEEE PRESS Marketing
Attn: Special Sales
PO Box 1331
445 Hoes Lane
Piscataway, NJ 08855-1331
Fax (908) 981-8062

Printed in the United States of America

10 9 8 7 6 5 4 3 2

ISBN 0-7803-0433-0
IEEE Order Number: PC0330-1

Library of Congress Cataloging-in-Publication Data

Lucky, R. W.
Lucky strikes — again / Robert W. Lucky
p. cm.
ISBN 0-7803-0433-0
1. Electrical engineering. I. Title.
TK159.L83 1993
621.3 — dc20

92-30769
CIP

Contents

Foreword

This is a serious book. Little of it is to be taken as absolute fact, yet there are strong allegorical messages. While you are reading it, you will enjoy it. Afterward, you will savor it.

You may wonder about the genesis of the essays contained in this volume. The way I remember it, the author and I were having a late dinner after some IEEE meeting back in 1981. Dr. Lucky (Bob to his colleagues) was IEEE Executive Vice President then, and I was Editor of *Spectrum* magazine. We got to reminiscing about how we each first became interested in electrical engineering, what our hobbies had been as kids growing up, where we went to school, and the like. Though our paths had not crossed until we became members of the IEEE Publications Board, we found, as engineers often do whose careers have taken different paths, that we nevertheless had much in common. Some of the anecdotes we shared stirred memories that had lain dormant for years. It turned out that Bob's professors at Purdue had similar idiosyncrasies to mine at Cornell, and that Bob had observed some odd but not inexplicable incidents at Bell Labs that bore an uncanny parallel to some odd but not inexplicable incidents I had witnessed while an engineer at CBS.

Suddenly Bob confessed to a long-simmering urge to recount in print some of the typical situations engineers encounter during their careers. They would be based on the same kinds of recollections we had been regaling one another with.

I thought it a great idea. My enthusiasm may have been abetted by the tenderness of the steak I had just consumed, or the conviviality of the occasion, or the fact that the meeting earlier that day was particularly successful, or even the unusually fine autumn weather.

In any case, I suggested a page in *Spectrum* as an experimental vehicle for a column that, by the time it first appeared, in January 1982, we had agreed would be called, simply, "Reflections."

The column became the closest of anything in *Spectrum*'s publishing history to being an instant success.

Bob's column runs every other month in *Spectrum* and has attracted a faithful following. Every so often we have a reader who, observing the column missing during an off month, demands its reinstatement. We have a form letter that guarantees we'll reinstate it the following month, and we have never failed to do so. We faithfully reinstate it six times a year.

Followers of the column find that while Bob's topics are often surprising, they rarely fail to touch a sympathetic chord. He is acutely aware of the engineer's dilemma: namely, that he or she is both a professional and a businessman. This dichotomous state is perhaps best defined by historian Edwin Layton in his classic book, *The Revolt of the Engineers.* Layton observes that in the United States the role of the engineer has been "a patchwork of compromises between professionalism and organizational loyalty," adding that "on the job he is part hired employee, part professional" and noting as just one example of conflict that "the professional may wish to publish the results of his work, while the employee may prefer to keep them as trade secrets."

Such conflicts are a recurring theme in Bob's essays. He takes good-natured gibes at lawyers, financial executives, and other corporate bureaucrats. In his post as Executive Director of the Communications Sciences Division at Bell Laboratories, Bob epitomizes the insoluble conflicts of bureaucrat and professional. From that vantage point he offers not-so-subtle advice to the professional on ways to circumvent the bureaucracy and, with an even hand, gives the bureaucrats some hints on when to look the other way.

He plays both sides of the street with aplomb and good humor. Would that more executives could recognize irreconcilable conflict and see the irony in it, and that engineers could be aware of their own reactive foibles.

Through many of his essays, Bob helps his colleagues gain perspective on the rampant pace of technology. As established technologies are displaced, he eases the pain with nostalgic remembrances of the importance they once held in the lives of engineers.

At *Spectrum* I always look forward to receiving the draft of Bob's column. It seems invariably in sympathy with my own experiences. I often skip the title sheet and jump right into the column itself. One night I awoke with what I believed to be an ideal — if a bit unusual — title for the column I had received the day before. It turned out to be the same title Bob had put on it. When I received his first few contributions I used to think "This is so good I could have written it myself." It was only later that I began to realize that most readers feel the same way.

The bottom line of course is that none of us did write them. Bob did. And we thank him.

Donald Christiansen

Postscript: In addition to the material originally published in *IEEE Spectrum* under the title "Reflections," this volume contains several new, previously unpublished essays.

Preface

Every other month for the last ten years or so I have faced a deadline — my "Reflections" column for *IEEE Spectrum* comes due. What a great privilege having that deadline has been! Nothing else that I could have done for our engineering profession in the way of invention or scientific discovery could have brought me as much satisfaction, joy, and just plain fun as has writing these periodic few paragraphs. It has given me a feeling of being in touch with the larger community of engineers, of tapping into the collective feelings of engineers everywhere. Being able to collect these essays in this book is a milestone that I have looked forward to for many years.

My aim in these essays has been to combine a little fun with a small message about our profession. We are, as a group, fairly serious about our work and about ourselves. Too stuffy usually. Sometimes it helps to lighten up a bit, and that is what this book is all about. For example, in the previous paragraph I said that writing these essays had brought me more satisfaction than any invention or scientific discovery that I could have made. Right away we have an instance of gratuitous stuffiness. You would be perfectly right to ask, "Well, has he tried inventing anything lately?" That hurts, but it is the kind of little truth that I seek.

An engineer friend once told me how much he and his wife had enjoyed seeing me discussing technology on some television show. He hesitated, obviously considering how to frame his next comment. "I don't want you to take this wrong," he said, "but my wife said something about you that

I think is perceptive. She said, 'He says what you think.'" My friend seemed embarrassed at the comment, sensing that it had not come out quite right, but I knew exactly what his wife had meant, and I took it as the best kind of compliment. That is what I am trying to do in many of these essays; I am trying to say things that are obvious to all of us, but somehow seldom said forthrightly in print. I do not expect everyone to agree with everything that I have written, but the more people who shake their heads affirmatively, and say yes, that is how it is, the more successful I will have been.

Because I enjoy writing this kind of material, I have used the excuse of collecting these columns to write more of the same. Lamely disguised as introductions to segments of this book is the equivalent of about three dozen more columns. The format just gives me the opportunity to expand on some of the themes contained in my old columns. So often I have wanted to return to some subject that I have already covered in a previous column, but have felt reluctant to do so. I know that if I were submitting the new material as a technical paper, the reviewers would say, "The author has already covered this topic exhaustively in his previous paper, and should be discouraged from using the precious pages of this journal in such a wasteful manner." Reviewers talk this way. The operant word here is "exhaustively."

A preface is an awkward format at best. (Can you imagine a preface in a technical paper?) It does, however, give me a chance to answer the two questions I am asked most frequently about my column. People always want to know how long it takes me to write a column, and they ask how they can go about getting their own column.

I think the reason people ask how long it takes to write a column is simply because they are being nice and want to make some comment, and this is the simplest question that comes to mind. For the record, it takes me about three or four hours. Possibly that seems fast, but consider if I had to do a newspaper column every day, I would be in big trouble. I don't know how the real columnists do it.

An associate who writes a column in another technical magazine once gave me a call about getting together to share notes about how we write our columns. I hesitated, uncertain how to answer. He went ahead, telling me about a file drawer that he keeps with ideas and partially written columns. I was embarrassed to confess to him that I had no technique at all. In fact, I had never previously thought about the necessity of having a technique. After that humbling experience, I started a computer file of subjects for future columns. The file is named, simply, "ideas." I have been adding to the file for a half dozen years or so, and there are perhaps a hundred possible subjects listed there. I have never used a single one of them.

My technique, such as it is, is as follows. The first step is to wait for the deadline. When the deadline has expired, I have sufficient motivation to spring into action. I sit in front of my computer, call in the editor (emacs

— thank you, Richard Stallman), and wait for inspiration. It is literally true that I have no idea of what the column will be about when I sit in front of the keyboard. The stale ideas in my computer file do not stir my imagination at the moment that they are needed.

Sometimes I have to start typing even when I haven't the foggiest idea of what I will say. Let's start this one with a *B*, I say to myself. Then I need a vowel, and from there on it is downhill. It is almost that bad, but something has always come to me. At least until now. I live in fear that the next time no inspiration will be forthcoming. For ten years this infallible technique has worked, but that does not mean it will continue to work. Check the next odd-month *Spectrum* to see if "Reflections" is still there.

So much for the easy question. The more difficult question is how does one become a columnist? People who ask this question usually have a particular columnist in mind — themselves. They look me in the eye, and I can see the little wheels in there turning. "I could do this better than you," is the printout going across their forehead. Trouble is, they might very well be right. I can sense their frustration. "How come you've got a column and I don't?" is what they are asking themselves. I always feel embarrassed and apologetic.

People have sent me example columns. Some of them were quite good, and I can only suggest that they send them to some editor, preferably not mine. I think that columns are hard enough to write without the added burden of having to ward off would-be competitors. However, I do have some sympathy for these people. If you write a paper, you can submit it to a journal. If you write a book, you can submit it to a publisher. But how do you get a column published? Alas, I haven't the faintest idea. Perhaps I should say that you have to be lucky, but people would think that I mean "Lucky."

Certainly it helps to know an editor, and without Don Christiansen these columns and this book would not exist. He discovered me, and I am most grateful. He has done even more than that; he has personally copyedited my submissions and drawn the cartoons that have often accompanied the published versions.

Unfortunately, this information does not help anyone else. I have no idea how to go about being discovered. Perhaps my editor even invented me, for all I know. Every time I face that bimonthly deadline I tell myself how fortunate I am to be able to face the blank screen on my computer, and to be able to fill it with random, sometimes meaningless, characters that so many people will eventually read.

So much for the preface. Thanks for reading this, and I hope you enjoy this collection of essays about our profession.

Robert W. Lucky

Part 1

About Engineers

Last week I did it again. The security officer at the Amsterdam airport asked me my occupation. "I'm a scientist," I said. Glancing through my passport at all the immigration stamps, she looked at me suspiciously. "Why does a scientist travel so much?" she asked. I hesitated momentarily. "Well, actually I'm an engineer," I confessed guiltily.

The security agent looked me over appraisingly. "Wait here," she said. "I have to get my supervisor."

While we waited silently for her supervisor to appear, I felt the weight of her scrutiny. I imagined that she was wondering what kind of person would change their story about what occupation they were in. I was less concerned about what she thought than about my own denial of engineering. Every now and then I do that almost unconsciously, apparently trying to give myself a classier label. Not that there is anything to choose from, of course. In the media, scientists are portrayed as evil and insane, but brilliant, whereas engineers are shown as nerds, but useful. Take your choice.

The Satisfaction in Engineering

I am ambivalent about being an engineer, and I think this is well demonstrated in the columns collected in this section. At its best, I have a great pride in engineering, and in being an engineer. Some of this is expressed in "The Joy of Engineering," where I recount some of the good parts of our work, in particular the satisfaction in the solving of problems, large and small, and in doing work that is ultimately of demonstrable benefit to society.

In the course of my career I have had many joyful moments as an engineer. Like others, I have traveled widely, written books and articles, made professional friends the world over, and celebrated promotions and awards. But the greatest joys that I have had as an engineer have been on those few occasions when I have had a genuinely good idea. There is nothing like a world-class idea to stimulate the greatest feelings of soaring euphoria, and by the term "world-class" I do not necessarily mean something that a newspaper would consider newsworthy, merely some little idea that maybe had not previously occurred to anyone else. An idea all my own. There is nothing like it.

Before I get all wrapped up in the glory of ideas, I should recount a story about Einstein. It seems that he was once approached by a student at Princeton who told Einstein that he kept a notepad and pencil next to his bed, in case he got an idea while he was sleeping. Did Einstein do something similar? Einstein is said to have shaken his head sadly and said, "I don't get ideas very often."

There is, however, a considerable satisfaction in everyday, little ideas — ideas that surmount the inevitable obstacles that nature has placed in our paths. I often think that the essence of being an engineer is being a problem-solver. Sometimes when I see people on airplanes working crossword puzzles I feel a sense of kinship. I don't work crossword puzzles myself, but I know the same feeling of elation that comes when I discover the bug in my little computer program. Maybe it is not "the thrill of victory and the agony of defeat," but these are the daily rewards of engineering.

There is a motto for a company whose name I have conveniently forgotten, but it says that "we bring good things to life." I hope they will not mind my borrowing their motto, because I think it fits engineers well. When I browse through a store and see something like a compact disc player, I feel a certain vicarious pride. We did that, I think. I want to grab some onlooker and tell them about the beautiful technology inside that cheap little player. Do they know about Reed-Solomon codes, or the semiconductor laser, or the sub-micron tracking loops, or optical disc storage? These are the objets d'art in our world. I look at the unappreciative patrons and I think that perhaps we should not allow these unwashed neanderthals to purchase our masterpieces — like the art gallery that puts "SOLD" stickers on the works reserved for the artists themselves. Yes, we do bring good things to life, and I am proud of it.

I had been considering these positive aspects of the engineering profession while waiting for the security supervisor at the airport. How could I deny such a noble profession? Now the supervisor started looking through my passport himself. "Do you know anyone in Italy?" he asked. I assured him that I did, thinking of my professional friends there. Seeing the stamps from Ben Gurion Airport in Israel, he asked me the same question about that country. Again I replied with pride that of course I did. Then coming across the stamp from Greece, the supervisor asked me about friends there. I was happy to admit that yes, I did know people there. Surely the supervisor could see that being an engineer meant that you were part of a worldwide fraternity of people with a common culture and language. But perhaps his mind was tuned to a different wavelength. "We need to see more personal documentation and business correspondence," he said to me skeptically. I had been caught up in bragging about the globalization of our profession, but now I realized that these had been the wrong answers to be giving to a security agent.

The Public Perception of Engineers

The episode with the security agent at the airport was another instance of what I think of as a "moment of truth" as described in "The Image Gap." It usually happens at a cocktail party when someone asks you what you do for a living. Intellectually, we can tell ourselves about the kinds of good points I have just mentioned, but emotionally we have a problem. Our public image is the nerd, and I never feel it so strongly as when I have to confess to some stranger that I am actually ... I stammer ... an engineer.

At the time that I am writing this section there have been riots in Los Angeles, and the police there have come under severe public criticism, first for exhibiting wanton brutality, and later for failing to act promptly when the rioting began. There was an article in the paper about a Los Angeles cop who said that things had gotten so bad that at parties he told people that he was a chemical engineer. Of course, I got a kick out of this. We all know that he could easily get away with this because no one would ask him anything about chemical engineering. I wouldn't even try myself. On the other hand, it has often tempted me at some party, when someone asks me what I do, to say that I'm a cop. Unfortunately, then people would be interested in me, and I could never fake it. It is awful to be so boring.

I mentioned earlier the little triumphs that come from solving problems. Often when I discover some little thing or uncover some particularly dastardly bug I clap my hands for myself. Then I instinctively look around for someone to whom I can explain my triumph, and I always rediscover the bad part of engineering — there is no one to share the satisfaction of a job well done. I might as well clap for myself; no one else will.

Perhaps this isolation is the most difficult aspect of engineering. We cannot tell our spouses or our families what we do. None of our nonengineer friends has the slightest knowledge or interest about how we spend our days. Unfortunately, even among our colleagues there is often very little enthusiasm for hearing about our own particular projects. Furthermore, the increasing depth of specialization of many engineering disciplines takes us further and further from each other. Even so, I always feel much more comfortable among engineers than I do among stock brokers, lawyers, or insurance agents, for example. I often think about this sense of kinship with engineers and scientists, but I cannot completely explain it. We seem statistically to have common backgrounds, cultures, and outside interests. I suppose it is all part of what makes us engineers and scientists in the first place.

Because of the difficulty of our work, engineers are never portrayed with any degree of realism in public. One of my most common fantasies, recurrent in these columns, is a television show about engineers. Norman Augustine has suggested "L.A. Engineer." The idea itself is laughable, yet I always feel envious of doctors, lawyers, detectives, and other television occupations. Imagine being able to tell people what you do, and having them actually be interested! Yet it has been pointed out that there is not a single person portrayed on television who makes a product or who creates actual wealth. Everyone on television is in the service business, and they deal with people, not things. It is a plastic world — no engineers, thanks.

I said that we seem to have a lot in common in our interests. This is also a public perception. They largely perceive engineers as useful but, well, boring. We are seen as one-dimensional people, with little interest in the social world that engages everyone else. Someone has to work with these boring machines, so it is just as well that there are people with such twisted interests — so long as they do not contaminate normal people.

For myself, at least, it is true that I have a fascination with gadgets. I love computer fairs, electronics flea markets, and consumer electronics stores. I remember the keen visceral thrill I experienced when I first visited the multi-story electronics store strategically positioned just at the gate of a U.S. Air Force base in Japan. There it was, lying in wait for someone such as me. And apparently there were a lot of me, because the store was full of people completely ignoring each other because of their rapt attention upon the unbelievable goodies displayed on shelf after shelf. Surely this was engineer heaven! On my way back from that trip I wrote the column "The High-Tech Appeal."

It is true that we engineers and scientists have a special relationship with things and concepts. For example, we alone hold the special secret knowledge of how to program VCRs to record television shows. Nonetheless, I keep insisting to myself (no one else will listen!) that engineers have other talents, that the one-dimensional caricature is surely unfair. I have known many engineers who were fine musicians, athletes, or had deep appreciation for art and literature. (I have known very few engineer-artists, however.) Yet it seems that people assume that engineers are innately incapable of doing anything except caring for machines. Periodically someone will say to me that I write well, or speak well, or do whatever well — "for an engineer." The qualification in that praise always gets my attention. I can never decide whether to take it as a compliment or an insult, but I am inclined toward the latter.

Some years ago I was in an IEEE committee meeting in which the results of a recently completed survey of engineers were shown. I remember commenting on one category in which engineers differed dramatically from the rest of the population — their divorce rate was extraordinarily low. One woman at the meeting muttered under her breath in disgust, "They haven't got the imagination for it."

Not only do we engineers seem to have common interests, but sometimes I hold the suspicion that we even look alike. Years ago I worked in a laboratory that developed data communications equipment. Some of the engineers in that lab used to joke with me that there was a "data-sized" person — that all the male engineers in the laboratory had the same general height and build. Even now when I run into some of those people they joke about old so-and-so, who just couldn't make it because he wasn't "data-sized." There is always an undercurrent of nervousness in our laughter, however. You don't suppose there is anything to this, do you?

So what does a typical engineer look like? Well, he is not Robert Redford, and she is not Elizabeth Taylor. We have to set our sights considerably lower, according to the little poll I took in "Celebrities as Engineers." If you want to get depressed, read it and look in a mirror. Hopefully you are atypical! For myself, there may be some glimmer of hope as old age sets in, because the other day I was being met at a small airport by an engineering faculty member whom I did not know. As I came in with a dozen or so other passengers I saw him waiting at the gate. I recognized him immediately; he looked like an engineer. I looked directly at him, but his eyes were resolutely focused on the man behind me. I tried to intercept him, but he rushed up to the stranger, saying "You must be Dr. Lucky." I studied this impostor closely. Yes, he too looked like an engineer.

Prestige and Pay for Engineering

I confess to a certain envy of the public perception accorded the medical profession. People seem to think that every doctor is brilliant in everything, whereas an engineer who may have been trained an equal amount of time is regarded as one-dimensional. The doctors are very good about promulgating this image of omnipotence. You can be standing beside an MD at some party, and someone has some technical question, and they turn to the doctor for an answer. Wait a minute, you think, *I'm* the technical expert. But you dare not say anything, because you know the immediate response, spoken or unspoken, would be, "But he's (or she's) the *doctor.*"

Many engineers who have worked with medical people in the biomedical engineering field tell stories of how they have been treated as second class citizens. I had my own sort of experience not long ago when I had a tennis injury to my shoulder. My secretary made an appointment with a local orthopedic physician. In a ruse to get me better service, and unknown to me, she made the appointment in the name of "Dr. Lucky."

When I arrived at the doctor's office, there was a waiting room full of people, but I was immediately ushered into an examination room. It was only a few minutes later that the doctor himself rushed in, holding the ubiquitous clipboard in his hand. "I saw you were a doctor," he said breathlessly, "and I didn't want to keep you waiting." Foolishly, I blurted out, "Oh, not that kind of doctor — I'm an engineer." The doctor looked up at me in astonishment, abruptly turned his back, and walked out of the room without saying a word. I didn't see him for another hour. Oh the ignominy of it! Like our time is not worth anything.

Anyway, as I said, there is joy in engineering — but what about good pay? I suppose it is all in how you look at it. Typically, engineering salaries for recent graduates top the charts. It is also relatively easy to get a job in comparison with graduates in liberal arts and other academic disciplines. Unfortunately, those salaries that look so good coming out of school often grow only slowly during an engineer's career. Someone who stays in engineering may well find that their salary gets outstripped by people who they remember as being only mediocre college students.

Nobody ever guaranteed that life would be fair, but I often find myself envious of business school graduates who go on to wheel in big bucks almost as if it were their due. During the "greedy" decade of the 1980s, we had a phenomenon where some of the top students from electrical engineering went immediately on to business school and thence to Wall Street. One dean of engineering told me that at a college theater showing the movie *Wall Street,* the students were cheering for Geckko, the corporate raider played by Michael Douglas.

It is not that engineers do not get rich. Several of the most wealthy people in the United States are electrical engineers. For the most part, however, these people got rich as entrepreneurs, not from doing the engineering itself. Either way I have no complaint whatsoever; these people have created ideas, products, and jobs. I confess that my anguish concerns people in the investment community who get quite rich by moving money from one place to another, and taking a percentage as the money passes by. Often it is the engineer's work that is being bought and sold. Although the financial people perform a valuable function, the question is the relative degree of the reward.

I often ponder the inequities (as I see them, of course) of the way that society rewards various occupations. Some days I feel that it is stealing money to be paid for our kind of work. Then other days I see young, mediocre baseball players holding out for salaries in the millions of dollars. Why so much? What is the market feedback system that settles on such amounts? In "Layers of Abilities" I consider the skills of athletes and engineers, and conclude that there are orders of magnitude more variation in the skills of baseball players, for example, than in the skills of engineers. Put in its best light, it means that we are all pretty good, and I often sincerely feel this way. However, it also means a dearth of superstars and a relatively small spread in our salaries. If you do not care for such democracy, then start practicing your batting!

The Supply of Engineers

For as long as I can remember, I have been seeing reports about an imminent shortage of engineers. I have often sat in committee and board meetings where such reports have been discussed. In this kind of an august setting, it is easy to get carried away with your own importance and with the gravity of the situation. The country needs us, you think. You look around the table at the grim, determined expressions of your fellow members. There seems to be a kind of competition as to who can look most visibly concerned. After all, the well-being of the entire free world is at stake. There must be a way to attract more students into engineering. People trade suggestions about how the seriousness of the situation can be brought to the attention of government.

Leaving meetings like this I sometimes feel the disequilibrium of a return to the real world. Was what we were doing really so important? Or were we only trying to play up our own importance? In the particular case of predicting the need for engineers there is a strong bias toward erring on the side of a shortage. In fact, I can never remember seeing a report that said there would be too many engineers in the future. After all, since we are so good, there should be more like us.

Back at my hotel, I consider the credibility of the report on the shortage of engineers. Who is going to believe us? Would we believe a study conducted by lawyers that concluded that the nation faced a crisis owing to a future lack of lawyers? I suppose I endured this cold realization once too often, so I was determined to write a column about these kinds of reports. The result was "Engineers: Dearth or Glut?" In writing this column I had the familiar problem of having good friends on both sides of this controversial issue. Not wanting to offend good people, I tried to keep the tone of the column relatively neutral. I am afraid my own biases show through, however. See what you think.

Who Is an Engineer?

How many people choose engineering is one side of the issue; the other is what kind of people choose engineering. In the early 1980s there was a simple answer as far as electrical engineering was concerned — smart people, or perhaps we should say people with good high school grades. In that period the supply of engineers was not determined by how many students wanted to pursue engineering, but by the capacity of the university system to handle engineering students. There was no need to encourage anyone to choose electrical engineering, and in fact many universities were deciding upon implicit and explicit methods of discouraging students from that choice.

Many of my electrical engineering friends felt a certain small guilt about the fact that they themselves would not have been qualified for their present careers, based on the criteria being used by their universities. It is a disquieting thought and, moreover, it is not obvious to me that any profession or occupation should consist solely of the people with the highest grade point averages. This was one of the worries that prompted me to write "The Best and the Brightest."

In retrospect, this column from July, 1985, may have represented a high water mark in the attractiveness of electrical engineering. Since that time the demand has shrunk, and the grade point average of incoming students, while still quite high, has declined. In 1992 the electrical engineering enrollment in many engineering schools has slipped behind the enrollment in mechanical engineering. I hope my ME friends will forgive me, but I keep asking myself why anyone would want to be an ME when they could be an EE? Unfortunately, I can think of an answer to my own question, and that is that EE is not as much fun as it used to be. Heathkit has even quit making electrical kits, which is heresy to any old-time EE. Meanwhile, mechanical engineering has discovered computers and CAD, and might expect to get more students, assuming conservation of the total number of students interested in engineering in the first place.

Like many engineers, when I was in high school I did well in math and science. My teachers told me that I was an engineer, which was news to me, because I hadn't the faintest idea what engineers did. It was not as if I sat down one day and decided rationally among professions; it was just taken as a foregone conclusion that I was destined to be an engineer. Consequently, I have often felt that I was born an engineer. It is only in recent years that I have asked some of my engineering friends about their own career decisions and have discovered that many have similar stories, though I could not find anyone else who would admit to being born an engineer — terrible thought that that is!

My ruminations about how engineers come to be engineers led me to write the column "Engineers: Born or Made?" One of its main points is the thought that in the future, engineers will have to be found among people who do not instinctively consider themselves born as engineers, people who do not blitz their way through high school math and science but nonetheless have much to offer our profession through their diverse mix of skills. We are in danger of becoming a boring old occupation anyway, and I look forward to the new regime.

Engineering Education

Bound up in this question of who becomes an electrical engineer is the interwoven question: just what is an electrical engineer anyway? Earlier in my career I thought I knew the answer to this question, but I have since lost the faith. Is it someone who does circuit design? How about device fabrication or software generation? And what do all these people have in common? It seems that our profession is built on shifting sands, and the direction of the wind keeps changing. Meanwhile, we are blessed with an academic tradition known as tenure, which inserts a large flywheel into the academic definition of what an engineer is and what he or she should know.

Like many of my columns, "The Curriculum Dilemma" had its genesis in a real event. This one came from sitting next to a stranger on an airplane who was writing a defense of the inclusion of his own speciality — electromagnetic theory — in the electrical engineering core curriculum. I do not mean to imply that EM Theory should not be in the core curriculum; I simply do not know any more. There is only so much time — though some educators are pushing for a five-year degree — and there is too much to know. There are also too many different specializations that are spreading ever further apart. Maybe there is no more true EE anymore. And there is always the possibility that I don't know what I'm talking about.

Whatever you might think about the usefulness of electromagnetic theory to the average EE of today, it has been a part of the ritual of becoming an electrical engineer for a long time. If our profession has some of the elements of a fraternity, then part of the initiation that we all shared was learning Maxwell's equations. I have seen T-shirts with Maxwell's equations on them, and I have always smiled. Yes, I learned them too, and I am proud of us.

One of the other beautiful aspects of dear Maxwell's legacy is that his equations are really rather difficult. We are sorry, but these are not intended for consumption by the masses. After all, we have to keep something to ourselves, because so much of what we do today has become popularized in the magazines and newspapers. It was seeing all those computer journals in the newsstands that prompted me to write "Giving Away the Store," in which I worry about how everyone seems to know about computer technology these days. At the same time that the state of the art in electronics descends into metaphysical levels of specialization, the accessible upper layers — the computer applications — have become commodities for the masses.

I always feel a certain sense of diminishment when lawyer friends, or whatever, start discussing with me their favorite word processor programs, the intricacies of databases, or things that they are doing with computer networks. I think that I can see a certain gleam in their eyes that says that maybe what I do is not so difficult after all. I am always tempted to give an impassioned defense of engineering. We do a lot more than that, I want to say. Have you seen Maxwell's equations, for example?

Now, though, I have come full circle. Not many paragraphs ago I was lamenting the fact that the public has no sense of what engineers do, that we cannot discuss our work with our families. But when there is an area — computers and computer applications — that suddenly becomes public knowledge, I feel a loss of esteem. It is as if a hole has opened up in the middle of our profession and all kinds of knowledge is pouring through into the public domain. So I feel ambivalent. I want the exclusive knowledge and public respect, but I also want "L.A. Engineer." Whether or not we can have it both ways, we will have to see where technology leads us.

The Joy of Engineering

Don't you just love the little snippets of conversation you overhear in public places? Often, lacking context, you are left mystified. At other times a few words carry volumes of meaning. As I was walking in an affluent district of Washington, D.C., two obviously well-to-do men approached me on the sidewalk. They were dressed in three-piece suits and were carrying custom leather briefcases. As they passed me, one was saying laughingly to the other, "Would you believe that I started out as an engineer...?" Leaving me in the wake of expensive aftershave, they hurried about their business of being important. I decided that the one who had started as an engineer recovered from this dread disease before suffering permanent financial disfigurement.

As for myself, I have no regrets at being an engineer. I even get these moments of revelation now and then that crystallize the feeling. I'll be in some meeting or discussion about technology, sipping hot coffee in a reflective mood. Out of nowhere the thought recurs, "I really love this stuff." I try to maintain a neutral expression while this maverick thought finishes its interrupt routine in my head. After all, I'm paid a decent amount of money for this work, and it wouldn't do to show any undue enthusiasm. (This technology stuff is hard work, you understand.) But let's keep it a secret — there can be real joy in engineering.

Of course I realize that technology alone doesn't always suffice to keep bread on the family table. Most of our jobs involve a lot of other tasks and responsibilities, such as interfacing with people, paper, and bureaucratic systems. Alas, there is not always joy in our Mudville either, but most of us enjoy the technology content of our work. We enjoy solving puzzles, creating order out of disorder, and uncovering underlying principles. We appreciate the beauty of mathematical models, and above all, we have a curiosity about how things work. Those of us who don't — well, perhaps we move on to the three-piece suit, custom briefcase, and whatever. Whether that is upward or not is not always obvious.

Quite often I also reflect on how fortunate I am to be an engineer, and to be paid for doing such stimulating work. There are so many jobs that seem so undesirable; I'm sure each of you has your own list of occupations you'd never do. Ironically, there are probably many people in the world for whom engineering would be near the top of dislikes. Whenever I start to believe I understand how other people feel about their occupations I remember a time when I was on a radio station talk show discussing automation. I was saying that the more boring a job was, the more likely it was capable of being automated. Therefore, thought I, automation is good, since it rids the world of boring jobs and frees people for creative thought. Someone called into the radio station and said to me, "You know those boring jobs you were speaking about? Well I have one — and I like it. Who are you to decide what's good for people?" I'll try not to make that mistake again!

When I count the blessings of engineering as a profession I start with the fun of working in a creative and productive field that is always changing and expanding. I feel as if I'm at the center of an expanding universe. All points in the universe are moving away from me at incredible speed, representing knowledge in the ever-widening subspecialties. There is no way to keep up, but it is exciting just to be there. I drift this way or that, trying to chase passing fragments of fields that look important for the moment. The hopelessness of the overall chase is a fact I keep secret from myself, because every now and then I create a new little point in the universe myself — an idea or invention all my own. That doesn't happen very often, but it is surely the greatest thrill of our jobs, and it is a very special joy found in only a minority of occupations.

I'm proud that engineering is an international profession. Because of our common heritage and our publication system, I have friends all over the world. Engineers also do more of their share of traveling. I don't know the statistics, but I often observe that airplanes are full of engineers. Just look sometime at the reading material that travelers bring, or listen to those snippets of conversation while airborne. Furthermore, we get paid a reasonable if not princely salary, and we've been immune from unemployment — give or take a massive layoff or two!

Of course, not everything is perfect, and in my own little fantasies, I would change our profession in a few unimportant ways. First, I think I would get a new name for us. "Engineer" has too many connotations of steam locomotives and tough, manual labor; it is also too much a catchall term. Something like "certified electronics architect" or "computerologist" might convey a little more dignity. I'm kidding, of course, but occasionally I fail my own acid test of pride. The other day I told the customs inspector that I was a scientist. He couldn't have cared less, and I was ashamed of denying my own profession. I am also often bothered by the isolation of our work. The specialization of our fields makes it almost impossible to talk to other engineers, let alone laypeople, about our work. Do you talk to friends at cocktail parties about your work? I don't know how to change this isolation, but somehow I dream of a world in which there can be a hit Broadway musical based on the life of an engineer. (Dream on, O naive engineer!)

Obviously there are richer occupations. Others are closer to the flow of money. They have their hands in the stream as it passes by. Enough sticks to their fingers that it seems the reward is incommensurate with the contribution. Although some of the wealthier people in the country are electrical engineers, generally speaking we are not rich. And maybe on some scale we aren't as important as those unknown passersby with their custom briefcases. But there is always the joy of engineering.

The Image Gap

You're standing at a cocktail party and someone asks what you do for a living. It's the moment of truth. Do you say "I'm an electrical engineer?" Or do you cop out? "I'm in management," or "research," or "electronics" seems vague enough to evade the issue. The worst is to give in to the secret impulse to say "I'm a scientist." It's like when I'm in California and someone asks me where I'm from and I reply "from the New York area." When asked to be more specific, I have to admit sheepishly that I'm from New Jersey. There's a certain image problem.

William F. Buckley Jr.'s spy hero Blackford Oakes is an engineer. In *Saving the Queen* he explains that this is a very useful career — when he is asked his profession at a party, he need only mention engineering in order to avoid further conversation. ("Both ladies asked the usual questions and Black reestablished, for the thousandth time, that one had only to say one was an engineer to catalyze that haze-in-the-eye behind which all attention wanders.") I've seen it happen all too often, and probably so have you. One of my more talented friends told me that when he worked at Bell Labs he had hit upon the solution of replying that he "worked for the phone company." Assuming that he climbed telephone poles or installed telephones, people were then very friendly to him.

Several years ago IEEE's Board of Directors decided to fund an expanded public-relations program. A consulting firm was called in for advice. I remember being told by its representative that their firm had been engaged years ago by the American Medical Association, and that it had conceived of the very first program about doctors on television. So, thanks to PR, the doctors made it big on TV. The image of a show with engineers for heroes dangled implicitly before our eyes. I had visions of Alan Alda playing a design specialist, engagingly bedeviled by glitches in the multiplex timing of his memory circuitry. Or perhaps we would feature a software person — played with faint Shakespearean overtones by Richard Chamberlain — whose dramatic confrontation with a bug in his error trap routines changes his life. Ah well, it might as well have been sugarplum fairies. Lawyers, doctors, teachers, detectives, and business tycoons make good TV fare, but let's face it, would even we watch an engineer show?

On the other hand, I suppose we could outright buy a TV commercial time slot. I can see it now. We begin with an impressive-looking engineer in a three-piece business suit walking confidently toward the camera, holding a microprocessor chip in his hand and softly humming a song. (Naturally, he is really an actor. At the bottom of the screen in small print it says "Engineers in this sequence are simulated.") Soon he is joined by another engineer carrying a small stereo-cassette player. Others follow, carrying progressively larger items — a portable TV set, a home computer. Another is pushing an oscilloscope; then a small group pushes a rack of electronic equipment. Finally there are a thousand well-dressed, benevolent-looking engineers pushing a nuclear power plant. (Another small caption says "Nuclear power plant simulated and harmless.") The chorus swells from a whisper to an ear-shattering volume, singing "Just look for the EE emblem...."

The problem is partly that the general public doesn't have an image of what engineers do, and partly that it doesn't care an awful lot, either. Personally, I was an engineer before I had much idea of what we do, and sometimes even now I have my doubts. Most professions have a kind of public image, usually fashioned around famous members. At a cocktail party every lawyer can be imagined as a Clarence Darrow or an F. Lee Bailey, arguing life-or-death situations before a breathless court. Or any scientist can be an Einstein (or today even a Sagan), pondering momentously the future of the universe. But, quick now, name a famous electrical engineer that you could be imagined as being.

The tragedy is that I happen to think that many engineering jobs are indeed exciting. Computers are fun, communications is where it's at, and power has a great current importance. Furthermore, I feel that engineers create societal good, instead of playing the parasitic role that I (jealously) believe characteristic of some other unnamed, richer professions. But the intricacies of our jobs are beyond the public's comprehension or care. Tracy Kidder's *The Soul of a New Machine*, the current book about a team of computer engineers at Data General Corp., contains a number of observations about this public image of our profession. "The computer's reputation for awesome obscurity and the real complexities of the engineers' trade made barriers that were hard to cross. Their wives, some of the team said — and some of the wives agreed — didn't know much about what they did all day." One engineer laments, "No one understands what we do." "They lived in a land of mists and mirrors," Mr. Kidder summarizes with an unengineerlike phraseology.

The public may lack an image of what an engineer's job is, but there is no such problem with identifying engineers themselves. The caricature must date back to antiquity, perhaps even as far as when I was in school. Kidder credits the Data General team with describing the typical engineer as "... an imaginary creature perhaps, who wears a white undershirt and a plastic pouch (known to some as a 'nerd pack') in his breast pocket, in order to keep his pens from soiling his clothes. An electronic calculator — used to be a slide rule — hangs like a ring of janitor's keys from his belt." Needless to say, there aren't a lot of these actual types in Data General, but, admits one engineer, "They exist. And they're the most obvious engineers, because in their isolation they're obvious."

I'm often impressed with the talents and abilities of my fellow engineers at the laboratory where I work. It seems as if we have people who excel at many forms of human endeavor — athletics, music, art, and other aesthetic pursuits. Yet the instant an engineer gains recognition for such activities, he ceases, by definition, to be an engineer. Catch-22. Have you ever been told that you do something well, "for an engineer"? What kind of a compliment is this?

Obviously, there is an image gap. But what to do about it? That expanded public relations program I spoke of earlier — it no longer exists. It didn't make a dent, so don't expect to see engineers in light beer commercials. Just go back to your computers and forget about the outside world. And the next time someone asks you at a party what you do, just tell them you're a shepherd.

The High-Tech Appeal

I am writing this column while sitting in the cargo compartment of a large military aircraft returning to the United States from Japan. Stacked about me are the suitcases, bags, and boxes of the passengers. Guess what type of baggage is most prevalent. Right — box after box of Japanese consumer electronic gear, the newest prize possessions of my fellow travelers. I feel pangs of jealousy shiver through me. I wish I had that stereo receiver I am staring at now. And there is a beautiful video cassette recorder to my left, and a magnificent computerized audio cassette unit underneath it. Oh, I covet them!

The irony is that I do not even need them. Like most of you, I already have similar gadgets in my home. But my gear is ancient — almost two years old. It doesn't gleam as I imagine the contents of those boxes do. It doesn't have the computerized, adaptive, programmable gizmo or the multicolored, intelligent, ever-changing readouts and displays. I could open an electronics store with all these seductive packages and spend every day basking in the warm glow of all those beautiful gadgets. Oh, I love technology!

Apparently I have much company in this perversion. Electronics stores are full of mostly young people caressing the buttons and knobs of these gorgeous boxes with wistful longing. (What did we ever do before we had these adult toys, I wonder.) It seems that high technology is an art form, perhaps akin to jewelry, representing somehow the worth of its owner.

Several years ago I was scheduled to entertain a Japanese visitor to my laboratory. Not feeling up to the physical demands of dealing with what I anticipated would be a difficult communications channel, I delegated the role of host to a corporate subordinate. Feeling guilty during the actual visit time, I strolled down the corridor past my associate's office. The Japanese visitor was standing to leave, bowing as is the custom of these polite people. My guilt intensified. Now the visitor reached into his briefcase and withdrew a beautifully wrapped present and handed it to my associate — a present, needless to say, which had been intended for me. My guilt was temporarily overtaken by my sense of loss.

My associate unwrapped the gift as I unobtrusively caught sight of the contents from the corridor. It was a hand-painted paper fan. Now such fans are lovely, and giving one is certainly a thoughtful gesture, but frankly I didn't need another Japanese fan. My sense of loss disappeared; the guilt returned.

My associate thoughtfully set the fan on his desk and contemplated it, doubtless, I thought, composing his acceptance speech. Suddenly, I perversely envisioned him pushing the fan back across his desk toward the visitor. "No," he says, "I want a hi-fi set." In my fantasy the Japanese stares at the fan with puzzlement. Reaching down, he pushes the fan across the desk toward my associate. "Fan?" he says tentatively. Decisively, my associate once again returns the fan to the visitor. "I want a hi-fi set," he reiterates. "Hi-fi set."

The Japanese rises to leave, his face a mask of puzzlement, the spread fan lying untouched on the desk, staring upward like some unmentionable flaw in one's food at a formal banquet. His face clears as he evidently comes to the conclusion that he had not properly understood the acceptance speech. Bowing once again, he takes his leave.

Yes, we do want hi-fi sets. We want calculators, and digital watches, and all that wonderful electronic gadgetry. And yes, my Japanese friends, you understand us well. You know we covet that high-tech look, and you do so well in satisfying our dreams. You have my congratulations. Furthermore I know that, were hi-fi's small enough and inexpensive enough, you would indeed bring them as gifts. (I'm just kidding, of course. My colleague and I really don't expect gifts. Besides, it's against company policy. A ball-point pen, maybe, if it's not worth more than 17 cents.)

Undoubtedly there have been studies of the marketing attractiveness of technology itself. I've often felt this compulsion in myself and seen it in others. There is this need to own and control this caged power. I've known so many people who buy the finest cameras, lenses, and accessories — and either never use them, or never take meaningful photographs. I have known people who buy the most expensive of hi-fi's and spend their time listening for the brilliance of their highs and the power of their lows, yet have no appreciation for the content and meaning of the music itself.

Now with the advent of the home computer a new pinnacle of technical adoration has been reached. In the early days of the hobby, when I built my first home-brew system, people would ask me what it was good for. I have been unwaveringly and uninterestingly honest on this score. I tell them that it is not good for anything. I just like *having* it. It makes me feel good. I like to make it do useless things and watch the results go by on the screen. I write systems and applications programs in every known language. None of them controls the temperature in my home; none of them balances my checkbook or keeps an inventory of kitchen supplies, or does my Christmas mailing list. Most of these uses are crazy anyway, but I do not *need* them in any event.

Why, I wonder, do computer owners feel the need to rationalize their ownership? Nobody asks stamp collectors what stamp collecting is good for. People with other hobbies are allowed to do useless things that take them out of their normal existence, but for some reason this is not so in home computing.

As you might guess, we engineers are very susceptible to the desire for ownership of technology. Several years ago *Spectrum* did a readership study for qualification of their advertising appeal. Guess what? We own all kinds of fancy cameras, hi-fi's, VCRs, digital watches, and things like that. We're way above the national average. And how often do you see an engineer with a featureless camera or a packaged stereo? No, it has to glitter. It has to have buttons and displays, and things you can control even if there is no associated function. It has to have that high-tech look, like those boxes in front of me now. If I stare at them any longer, my sanity will be overcome by my desire and I will break into one of those boxes.

Celebrities as Engineers

Last December the marketing folks at my company decided that it was time to make a high-tech television commercial. Cameras, lights, crew members, assorted directors, and Cliff Robertson descended on our labs. Someone found that the molecular-beam epitaxy machines made a perfect backdrop, probably owing to the dry ice clouds wafting about them in the best sci-fi tradition.

Everything was in perfect order. Only one small detail remained: some real engineers should be in the background running the machines to lend authenticity and the personal touch. When the call went through the building for engineers to appear in a television commercial, a long line quickly materialized.

The director examined each of the 50 or so engineers who presented themselves. With his hands he simulated television frames about the heads of the candidates. He rejected them all; none of them looked like engineers. They had to send out for actors to play the background engineers.

This commercial has played many times on television. If you happen to see it, look closely at the people in the background. They are, as my children would say, clueless. It is obvious to me that they have no notion of what they are doing. I don't think they look like engineers at all. The ad people tell me that they do, and that I have no conception of what the public thinks an engineer looks like.

I wonder. The other day I was met at an airport by a limousine driver who had never seen me before. I was worried about making the connection, but the driver walked right up to me, instantly picking me out of the several hundred people on the airplane. I asked how he was able to be so decisive. Perhaps he had been given a photo of me, I suggested. He said no, it was easy to pick out engineers. Nothing to it. I felt mildly offended. After all, he might have had the decency to be a little unsure. There was always the possibility that I might be a movie star or something.

These thoughts had run through my head on previous occasions, one of them being a communications conference several years ago in Seattle. It turned out that we shared the hotel with another meeting — an annual conference of funeral directors. On the first evening both conferences held cocktail parties on opposite sides of the same ballroom. From the mezzanine I surveyed the assembled crowds beneath me, wondering if I could tell the engineers from the funeral directors. I imagined a giant photo of the dual congregation on the cover of an IEEE magazine as a quiz, with a key on the back page showing the engineers in outline form. Are we obvious to ourselves? To the public?

The final straw is the recent ad for the movie *Revenge of the Nerds.* The ad features prominently a nerd. The nerd is wearing Bermuda shorts with argyle socks and wingtip shoes. His shirt pocket is filled with pencils and pens, and there is a pocket calculator strapped to his belt. The thick glasses he wears are topped with clip-on sunglasses. What academic discipline do you suppose these nerds have signed up for? (No, not pre-med.) I understand that the movie's nerds triumph in the end; they usually do. However, being on the losing side in this case looks awfully attractive.

I decided to run a small experiment to see how we engineers are viewed. I made up a list of about 50 celebrities, which I gave to some of the secretaries and administrative staff at the lab, who work everyday amidst thousands of engineers. They know what engineers look like. "I can pick them right out of the parking lot," one staffer said. On that basis, I asked them: Which of the celebrities on my list could conceivably be an engineer? And which ones, if you saw them walking down the aisles here, would astonish you by being engineers? In their case, you would conclude, the guards at the doors must have made a mistake — this person just does not belong here.

My list included such names as Robert Redford, Madonna, Gary Hart, George Bush, Jane Pauley, Albert Einstein, Jimmy Connors, Ronald Reagan, and James Bond, to name a few. How did the staffers categorize them?

My ego took a certain beating when I heard the derisive laughter that greeted the possibility of someone like Robert Redford being an engineer. One woman was telling another that when her brother picked her up the other day at the labs, he had commented that "these people [the engineers] look like they are from another planet."

Anyway, when I tabulated all the scores from my scientific experiment, there were a number of unanimous choices as nonengineers. Robert Redford, Paul Newman, Tom Cruise, and Elizabeth Taylor are too beautiful. Sorry about that. Madonna just doesn't fill the role, and Dolly Parton is too something or other. Joan Rivers got no votes, nor did Frank Sinatra. Grace Kelly was too regal, and Michael Jackson too spacey. No one was willing to give Ronald Reagan a single vote as an engineer. A quartet of brawny male stars fell a single vote short of achieving perfect nonengineerdom — Clint Eastwood, Sylvester Stallone, Charles Bronson, and Bruce Lee. (Whichever person voted for each of them as an engineer, thank you.)

On the other side of the ledger — the question of who among the celebrities might really be an engineer — pickings were rather slim. There was only one name on the list that everyone felt could be an engineer — Albert Einstein. Personally, I see him as a scientist, but if he is the way the public sees us, I take it as a compliment. Several political figures qualified as potential engineers. They included Bill Bradley, George Bush, maybe Gary Hart, Jimmy Carter (who *is* one), maybe Gerry Ford, and Mario Cuomo. In the sports world, Larry Bird made it, which greatly surprised me, as did Martina Navratilova, Billy Jean King, and Arthur Ashe. In the entertainment business only Bill Cosby, Jane Pauley, Gloria Steinem, Bryant Gumbel, Geraldo Rivera, and Connie Chung were seen as possible engineers. A shock to me was the close engineering victory won by Mikhail Baryshnikov. (You never know.)

More in the middle ground as nonengineers were Bruce Springsteen (nice try, Boss), Huey Lewis (who sings *It's Hip to Be Square*), Jane Fonda, Lionel Ritchie, Chris Evert, John McEnroe, Donald Trump (too rich), Richard Nixon, Dustin Hoffman, and Luciano Pavarotti.

To end on a happy note, there was quite a pleasant surprise: James Bond made it. Nerds of the world, unite! There is hope!

Layers of Ability

The major league baseball system in the United States has an interesting personnel policy called free agency. After fulfilling a contract of several years' duration with one team, a baseball player can turn free agent. His services then become the subject of a well-publicized auction among the various competing teams. Subsequently, even the most mediocre free agents sign for annual salaries in the millions of dollars.

Watching some of these wealthy ex-free agents playing baseball on television fills me with a jealous anguish. They may not be able to field or hit, but they can drive to the stadium in a Rolls Royce, whereas I, a reasonably accomplished engineer, have trouble keeping my old clunker in decent repair. Where is the justice, I ask?

In my fantasies I imagine a free agency system for engineers. When your contract expires with Acme Engineering, you turn free agent and offer your talents to the highest bidder. But, alas, even in my fantasy the million-dollar offers fail to materialize. The range of offers that any of us received would be quite narrow, with the awful possibility that our measly present salary would be at the upper end of the scale. This brief and shattering vision leaves me with the question of why mediocre baseball players are paid so much and the most talented engineers paid so little, relatively speaking.

Simple, you say. Supply and demand. There is a great demand for entertainment and a supply of only a few hundred baseball players at the major league level, whereas there are millions of engineers, all engineering at a more or less equivalent level. In the case of baseball, perhaps 50 million young boys (and girls) learn the game and aspire to stardom. Of that multitude, only about .001 percent make it to the major league level where they are able to earn their livelihood playing baseball. If we draw a curve of salary vs. percentile, there is zero salary for the first 99.999 percent of the relevant population. Of course, the last thousandth of a percent does very well. The curve is simply an impulse function at the right edge of the paper.

In engineering, by contrast, there is a much smaller pool of aspirants. (Too small, some would say, but that is another story.) Of those who enter engineering school, perhaps half actually become graduate engineers. The curve of salary vs. percentile is quite flat, with the best paid engineers earning perhaps four times the salary of beginners. (We exclude those wealthy aberrations who have left engineering for other occupations.) The good news is that there are jobs for everyone. The bad news is that there is so small a spread in salary that there may be little to which we might aspire. We have no major league, no free agents, and, quite possibly, no stars.

I wonder if the range of abilities among engineers reflects the small range of salaries. Is the best engineer only a little better than average, but a major league baseball player — however mediocre — almost infinitely better than the average amateur?

Some light on the latter part of this question was shed in a recent experience of an engineer friend. This engineer happens to be the best tennis player I know, having achieved the No. 1 ranking in the company's tennis ladder. One day not long ago he found himself with a free afternoon at a resort hotel in a foreign country. Since he takes his tennis racket with him wherever he goes, he put up a little notice on the hotel's bulletin board. "Advanced player looking for tennis match ...," he wrote, adding his name and room number.

Shortly thereafter he received a call about his notice. "I am ________," said the caller, giving a name familiar to us all as that of an aging but highly ranked tennis professional. He, too, was looking for a match that afternoon. "I didn't really mean 'advanced,' " stammered my friend. But in spite of his protestations of hapless mediocrity, the match was arranged — a match surely taken right out of our best Walter Mitty fantasies.

"How did it go?" I asked in eager anticipation. I could see myself out there, hacking away at the pro. I wondered how I would do. I mean, how good are they, really? With a small wave of his hand my friend deprecated the whole affair. What I had already heard was the best part. He said that he had played the match of his life, putting everything he had into every stroke, while the aging pro had paid little attention, sipping cocktails between points, and flirting with his courtside girlfriend. "So how did it go?" I persisted. "We played one set," he finally answered, "and I didn't get a single point."

I tend to classify activities based on the number of levels of skill that exist, where I think of a layer being distinguished by the ability of an average member of that layer to completely suppress someone from the layer below — consistently 6-0 in tennis, say. Perhaps there are six or seven layers in tennis, at a guess. How many layers are there between me and, say, Michael Jordan in basketball? Or between me and Karpov in chess, or Isaac Stern on the violin? Many, I am afraid — layers upon layers of ability. I remember a few years ago shooting a basketball in my driveway and making three or four dozen free throws consecutively. I was proud of myself, and out of curiosity checked the *Guinness Book of Records*. My pride quickly vanished in embarrassment; the record was somewhere in the thousands! You name it, somewhere in the world there are people who are unbelievably good at whatever it is.

Are there so many levels of ability in our profession of engineering? Perhaps — but perhaps not. The field itself is evolving at an incredible rate, and we all find ourselves riding the roller-coaster crest of a tidal wave of change. Some of us are swept under, and the rest of us swirl about in the foam, with our eyes fixed on the rocks below. I look ahead for the leading swimmers, the stars who are levels ahead of the pack, and I see an empty ocean. I look back for the laggards who have missed the waves and are merely bobbing about in a becalmed sea, and though I see some, that is not where the masses are. Most engineers are clustered on those crashing waves. I see competence, and even brilliance, wherever I go in our profession. I see it in individual engineers in small companies, in the large corporations, in government, and in academia. I am often amazed to find state-of-the-art engineering ability in the most out-of-the-way places.

It makes me proud to be a member of such a group. There is always room at the top, but the bottom is never far away either. Maybe we don't have stars who are levels ahead, and maybe we couldn't turn free agent, but I feel privileged to be one of such an outstanding company.

Engineers: Dearth or Glut?

Ever since I can remember, we engineers have been doing studies about the supply of engineers. Will there be enough of us in the future, we wonder? Invariably our duly constituted committees, panels, and commissions conclude that the country faces the dire prospect of a life-threatening shortage of engineers. In sombre tones of scientific gloom their reports foretell the future — we engineers are becoming an endangered species.

In the halls of government these reports are received with a deeply concerned mood of shock and dismay. "They say there won't be enough of them," says one congressman to another in a hoarse, emotion-laden voice. The other congressman stares fixedly at a blank point in space, his poker expression marred only by a pronounced facial tic. "Does the President know?" he whispers as he glances nervously towards the half-opened door. The only reply is the slightest negative shake of his associate's head. Sensing an impending governmental crisis, the second congressman dares to utter the forbidden question. "When didn't the President know about this engineering shortage?" he asks.

The average citizen is equally concerned. His thoughts hover constantly round the question. How many engineers will be needed in the future? Will there be enough? It is hard to go about everyday life with these eternal questions unanswered. Meanwhile, in response to the authoritative reports on the matter, sales of bumper stickers with the slogan "I brake for engineers" flourish, and committees of citizens struggle with local programs to recycle engineers.

Occasionally I worry about this too. Should there be more like me, or fewer like me? I am ambivalent. Sometimes I think that it should be manifestly evident that engineering is one of humanity's highest aspirations. What a national tragedy that so few of our youth select this noble profession. The country needs us; the world needs us. The more like us, the better.

On the other hand, we all know of engineers out of work because of cutbacks and layoffs. A vocal segment of our community cries out that there is actually a glut — not a shortage — of engineers. The fewer of us the better, they say. Then there would be jobs for all, higher salaries, and more respect. The vision of rarity dances before my eyes. I imagine telling someone my occupation at a party and not getting that "another one of those" look, but instead hearing the person stutter in awe, "A wh... what?"

I wonder, too, about this whole process of determining the future demand for engineers. One method is to take some benchmark period in which times for engineers were good, and to use the number of engineers then as a means of measuring future demand. However it is done, it is hard to avoid making the assumption that there is a certain amount of engineering to do, and engineers will be the ones doing it.

Such an assumption should work well for forecasting the needs for doctors, for example. You can always count on a certain number of people getting sick, and only doctors are allowed to do the doctoring. This excludes, of course, the possibility of an outbreak of the plague, or — as a doctor once complained to me about conditions in his hometown of San Diego — the outbreak of a health epidemic.

In contrast to the relatively fixed need for doctors, the need for lawyers actally expands with the supply of lawyers. The more you have, the more you need. What a wonderful property!

Not only is the demand for engineers a soft number, but it encompasses a fuzzy set of job descriptions. Which jobs are suitable for engineers and which are not? As the demand fluctuates, we dip higher and lower into the job pool. When I see EEs running hot dog stands, as we did some years ago, then we have certainly gone far too low. If I see lawyers designing VLSI chips, then I will know that we have elevated our talents to the choicest positions. But I won't hold my breath on that.

Now how about the supply of engineers? Strangely, it is fairly predictable. How do a certain number of high school students each year know to choose engineering? The world works in mysterious ways. We may not breed people to portion out to the occupations, as in *Brave New World*, but somehow a relatively constant fraction of students elect to be engineers. At issue is the exact value of that engineering fraction as it drifts slowly from year to year.

Some people say that we should meddle with this fraction, to push it higher or lower, whatever might be our wont. We do exercise some control by raising and lowering entrance requirements, but larger variations are harder to implement. The lawyers found a way to increase their fraction with the television show "L.A. Law." Sometimes I imagine a similar show for engineers with, as in the case of "L.A. Law," a slightly sleazy, but incredibly handsome, group of men and women. But instead of dealing with messy divorce cases, our EEs will be engaged in writing fascinating computer programs. It should be a great hit.

Other people say there is no need to meddle with the system. The MARKET will prevail. It is as if there were a bid-and-asked quote for engineers written in the sky. I find it puzzling that high school students who barely know what engineers are have this innate sense of the market forces, but they do.

I suggest we approach this issue of supply and demand in typical engineering fashion. We should develop a mathematical model for the whole macroeconomic system, and run a simulation on a giant supercomputer. The outcome might show that we were short 137 engineers, or some such number, that would have to be recruited or coerced in the grade schools.

So little Johnny comes home from school in tears. "Mommy, they said I have to be an engineer," be blubbers. "Why me, Mommy? Why do I have to be an engineer when Billy gets to be a shepherd?"

His mother looks at him with a firm smile. "Shush, Johnny, they know what's good for you."

We do, don't we?

The Best and the Brightest

I think that the home computer is at fault. Or perhaps it is movies like *Wargames*, or all those employment ads in the newspapers. For whatever reason, it seems that every teenager wants to be an electrical or computer engineer. Do you have a son or daughter graduating from high school who wants to be an EE? If so, welcome to the crowd. Does he or she have dreams of admission to the best schools? If you are blessed with a valedictorian, there is a chance. Otherwise, forget it. Deluged with applications, the schools have their choice of the cream of the crop. The best and the brightest of the next generation are headed into electrical engineering.

In the last few years, there have been persistent rumors that in certain schools the high school grade-point average for the incoming electrical engineering class was 4.0 (out of 4.0). A friend who teaches EE in a major university did a small survey of his associates on the EE staff. He concluded that none of the present faculty would have been admitted to the unversity's current freshman class on the basis of high school records. An even more amazing statistic at another major school was that the incoming EE class had a higher average *verbal* Scholastic Aptitude Test score than the incoming arts majors — not to speak of the disparity in math scores!

Schools where engineering students choose individual majors after the freshman year are faced with another problem — everyone chooses

EE. University administrators are faced with a disproportionate division of their resources. Confronted with tenured faculty and settled space in the schools being shunned, the administrators rationalize their inaction behind the pendulum theory of history. Meanwhile, the EE students have to deal with giant class sizes and "discouragement" from beleaguered faculty who take matters into their own hands. Students who are thus discouraged transfer to the less popular disciplines, and eventually everyone is happy. Only the best and the brightest remain.

Whenever I see movie clips of old football or basketball games, I find it hard to believe that the slow-motion clowns reenacting these events on the screen were once our sports heroes. Their modern counterparts dazzle in contrast. We have managed to breed a special mutant person optimized for, say, football linebacker. He is 6 foot 5 inches tall, weighs 260 pounds, and runs the 40-yard dash in 4.3 seconds. Any sport in which absolute records are kept confirms the dominance of the latest generation. The Olympic heroes popularized in the movie *Chariots of Fire* would not have placed in the average high school of today. What makes us think that electrical engineering is any different? Perhaps we have also bred a new generation optimized for electrical engineering. They may not be 6 foot 5, or weigh 260, but watch them tackle a keyboard or field an idea. Here they come, and they are the best!

Would it be possible to ask if this is all to the good? Is it good for the welfare of the world? Is it good for engineering? Is it good for all of us "experienced" engineers? Most probably the answer is yes, but let us consider for argument the contrary case. Imagine a profession peopled entirely by high school valedictorians. What image does this conjure up? Is it a profession to which you would wish to belong? Wouldn't you prefer to have a few dullards and misfits around just to break the monotony? Where are all the tradition-shattering innovations to come from? Think about all of our heroes — among them Franklin, Marconi, Henry, Faraday, and Westinghouse. How many of them would have made the grade? Couldn't we have just a few in the category of being voted "least likely to succeed"?

Most things in life are zero-sum games. If electrical engineering is drawing the best and the brightest, some field will be left with the worst and the dumbest. Maybe it will be the medical profession. Wouldn't that be just dandy: professionally we would compete against geniuses, while open-heart surgery would be done by the flunkouts. Or perhaps the U.S. Congress will become filled with people who never made the grade. (Did I say *will*?)

Obviously we cannot let this happen to our sister professions. It is up to us to take action. We must stop recruiting the best and the brightest and turn instead to the bottom of the class. Nobody with a grade average higher than C should be allowed into electrical engineering. Any spark of intelligence should call for immediate disqualification. This would of course require the development of very sophisticated SAT test methodology, lest shrewd achievers masquerade as know-nothings. Correlations between wrong answers would have to be carefully analyzed. Perhaps we would need a follow-up interview in order to probe carefully for any tell-tale gleam in the eye, indicating concealed intelligence. Probably such tests would have to be developed by nonelectrical engineers, since soon we would no longer be up to such intellectually demanding assignments.

There would be side benefits of the intellectual downgrading of EE. Difficult mathematics would gradually disappear from the textbooks. *IEEE Transactions* would become readable. Software would slowly become more transportable as clever and tricky coding became extinct. The human interface with computer systems would become more forgiving and humane. The onslaught of new chips would wane, and the obsolescence of computers would no longer be quite so critical a problem. We could probably do with half the number of publications and conferences. Engineers would lose their image of being nerds. Instead of playing the fumbling misfits and evil conspirators in the usual movies, we might even star in a beer ad or two.

Suppose we did it. Imagine that against tough competition we recruited a generation of academic losers. And suppose that the pace of technology continued unabated. What if engineering remained as abstract as ever? What if innovations continued to flow at a dizzying rate? What if the new generation started wearing thick glasses, forgetting their keys, and tripping over their front steps? Maybe it all has nothing to do with academic brilliance. Do we want to find that out? I think not. Forget this column.

Engineers: Born or Made?

People often ask me how I became an engineer, as if to say, how did you pick up this terrible disease? Since no one asks lawyers and business tycoons how they became what they are, I feel some resentment. Imagine someone asking, "Mr. Trump, how did you decide to become a real estate magnate?" I could imagine the Donald reflecting for a moment. "Well, first I considered engineering," he replies. "But I decided that the potential for philosophical gratification and self-realization in that, ah... profession, was not, well, what I had come to expect."

Usually I'm bored with this topic, so when someone asks that question, I simply say that I was born an engineer. That cuts the conversation short, with the questioner nodding in apparent understanding. I can see the thoughts flickering across the faces of these people. "Of course," they are thinking. "What a terrible burden to carry through his life. Amazing that he turned out so well — considering."

In one of my responsibilities at work, I often had the task of introducing military generals as banquet speakers. I was always given an official biography, but they all seemed the same to me. (I suppose our engineering biographies look the same to generals, too.) Something that caught my attention, though, was that these biographies always began with "General Whatshisname was born in Abilene, Kansas."

The concept of being born a general amused me. I would imagine the family and friends gathered around the baby's crib in Abilene. "Why just look at the little general!" they would exclaim. There would probably be three stars on the baby's diaper.

But instead, imagine the family that has just had a baby engineer bestowed upon them. As the expectant father paces nervously in the waiting room of the maternity ward, the swinging door opens and the doctor enters. He looks distracted, and avoids making eye contact with the anxious father. But as he hesitates in speaking, obviously trying to frame his words, the father leaps into the silence.

"Is it a boy?" he asks. There is a fractional pause — a kind of neutrality — in the doctor's silent expression. "Then it's a girl!" concludes the father breathlessly. By now the doctor has forgotten the comforting words that he had planned. There is the briefest negative shake of his head. "No, it's an engineer," he blurts out artlessly.

The father stiffens, resisting a display of disappointment. "You're ... you're sure?" he stammers. "I'm afraid so," says the doctor in his most practiced, soothing voice. Now a frown creases the father's face as an afterthought suddenly occurs to him. "Is it a boy engineer or a girl engineer?" he asks timidly. For a moment the doctor seems uncertain. "I'm pretty sure it is a boy engineer," he says reflectively.

Several moments of awkward silence pass before the new father probes the doctor's uncertainty. "How can you tell — I mean, about the engineer part?" he asks. The apprehension on his face seems to invite a withdrawal of the diagnosis. But the doctor shakes his head firmly to cut short this futile hope. "You can always tell," he says. "The normal baby appears with a contented expression on its face — almost as if it wants to burst into song — whereas your typical engineer ... well I don't mean to imply that there is anything *abnormal* about an engineer ... but they usually are frowning critically at the forceps. You can see their little minds ticking away, thinking about how they have to improve this place that they've come into."

Amused by this fantasy, I polled some of my engineer friends to find out how they became engineers. "Do you feel that you were born an engineer, or made into one?" I asked. No one admitted to being born an engineer. Instead they would give almost identical stories. "Well, I always liked to play with radios (or some such mechanical or electronic gadget)," they would say.

Then they invariably finished with a note of pride. "I was always good in math and science," they said to explain why. It seemed to me that their stories were tantamount to confessing to being born an engineer, but I suppose there are limits to self-revelation.

This matter bears on a problem our profession will face in the near future. Engineering will have to appeal more to women and minorities, and perhaps not everyone that the profession needs will feel that he or she was born an engineer.

One of my IEEE friends whispered a bit of heresy to me the other day. After glancing nervously around the room, he ventured an outrageous hypothesis. "Perhaps engineering is a trainable skill," he said. I composed my face into a shocked expression, and shook my head in proper admonishment.

Do you mean, I thought to myself, not daring to voice the words, that someone who hadn't been born an engineer — someone who hadn't played with mechanical and electronic things or hadn't necessarily aced the math and science courses — that such a person could actually be admitted to our sacred profession? What about our stiff entrance requirements and the series of tests and abstract mathematics courses that we have cleverly crafted for the early years of college? Only someone with innate engineering skills and total dedication to the profession could survive our weeding-out process. Only someone like us.

Sensing my disbelief, my IEEE friend hesitantly elaborated on his tenuous premise. "Suppose, just suppose, that we were to encourage people — maybe even nurture them — instead of just testing them?" he offered.

Well, this was going too far. Did this mean that a *normal* baby could actually become an engineer? My mind wandered back to the maternity ward, and now I imagined the doctor telling the new father that his wife has given birth to a healthy baby.

The father looks up in disappointment. "But we had hoped for an engineer," he says sadly. The doctor smiles, happy to be the bearer of good news. "You never know," he says. "Times have changed."

The Curriculum Dilemma

The man sitting beside me on the airplane was writing on a yellow pad — passionately, I thought, from the intense focus of concentration he was directing at the sprawling stream of inked characters he was producing so frantically. As soon as one page was finished, he ripped it off the pad and began another, his pen moving continuously. Now and then his hand rose and fell emphatically to hammer home an exclamation mark. Out of the corner of my eye, I could see that the script characters were getting larger and larger, as if the narrow lines on the tablet were insufficient to contain the frustration and energy in his writing. It was very easy to see he was angry.

I strained to read what he was writing without visibly turning my head. (Wouldn't you?) The few words I could make out had heightened my growing interest — something about Maxwell's equations. Now who could get so worked up about Maxwell's equations? Suddenly, I felt the thrill of secret knowledge. The poor fellow undoubtedly had no idea that he was sitting beside someone who was actually familiar with Maxwell's equations. Let's see, I said to myself, the divergence of B is zero, the curl of H is ... Well, I had known them once, anyway.

My head swivelled in his direction of its own accord. I had nothing to do with it; I never pry into other people's affairs. At this angle I could read his writing rather easily. After all, he had forced this upon me. Too bad I had missed the opening salvos, but now I could join this unfolding drama in the middle. From what I could make out, during a faculty curriculum committee meeting at one of our good engineering schools, someone had suggested that electromagnetic theory be dropped from the core curriculum. It appeared that mere words alone were not sufficient to convey the blasphemy of this untutored heathen's uncouth proposal. I could only conclude that this learned gentleman who sat beside me taught EM theory. Elementary, my dear Watson.

I risked a glance at the man's face. No, I had never seen him before, but then his face was so contorted with anger that maybe he was for the time being unrecognizable. Do away with EM theory? Never! I sat back and closed my eyes, and doubts began to percolate in my defenseless brain. There is a fundamental problem here that is not peculiar to EM theory. Knowledge in electrical engineering is growing exponentially, while the number of available course hours remains fixed. New things become important, and that means that from time to time old things have to go. But which old things?

Frankly, I do not know how we have gotten away with it for so many years already. There must be something that engineering graduates have been cheated out of learning because of the limited number of credit hours. What bothers me is that I am never able to identify this missing ingredient in their impressed knowledge. Surely my years of experience are good for something, I keep telling myself. I must know a multitude of things that they do not. But the graduates always smile knowingly; they know everything there is to know. The wisdom of the curriculum committee has sanctified their course work. Their degrees state that they are electrical engineers.

Even this mild thought is troubling. After all, just what is an electrical engineer today? What subject matter is the *sine qua non* of our profession? I used to think that it was circuit theory. Could you imagine an EE who had never studied Ohm's law? I still cannot, but I feel the foundations below me rumbling, and I worry. I see civil engineers and mechanical engineers programming their computers with knowing familiarity. I see physicists fabricating new optoelectronic devices and computer scientists studying communications protocols. The line has blurred, and not many of us worry from day to day about dear old Ohm and his sacred law of the universe.

So I understood the problem, and even with my eyes closed, I could empathize with my seatmate over the tension of curriculum pressure. But I had no answers either. Everything I knew about curricula was contained in a comment by an old professor that I had heard many, many years ago. He recounted how, when he had been a senior in electrical engineering, he had had a choice between two courses for his only elective. One course was in spark transmitters, while the other was in differential equations. "You know," he said wistfully, "I took the spark transmitter course." Although this story seemed intended to embody the wisdom of the ages, I have never been totally sure what to make of it. After all, the old professor seemed to have turned out all right anyway. Perhaps it only proves the principle of invariance that is sometimes whispered in the corridors of academe: it matters little what is taught as long as a certain engineering disposition is instilled. (Please do not pass this on.)

Whether or not it matters, the curriculum wars rage on. New concepts have been winning advocates, such as the suggestion to unjam the curriculum — that is, since there is no breathing room left, we should back off on the course requirements and free space in the curriculum for diversification. Or there is the proposed just-in-time curriculum, in which mathematical foundations are presented only as they are needed, instead of being loaded into the early years when the student has no idea of why he or she is learning these things. Perhaps a slight variation is the upside-down curriculum: have fun in the early years learning engineering applications, then do the hard work with abstract foundations in the later years.

Even the thought of these proposals, aimed at not dampening the ardor of the incoming engineering student, caused a virtual shiver to traverse my spine. Let them do it the traditional way, I decided; let them suffer. By all means they should study EM theory, preferably lots of it. It is good for them.

I opened my eyes and smiled at my seatmate. Go to it, I cheered him on in my mind. He looked over at me with a curious expression, as if I were the crazy one. Maybe he was right.

Giving Away the Store

Wisps of conversations drift over me here at 37,000 feet. "But the ground clutter will overwhelm the signal under those circumstances ...," comes from up ahead of me. Behind me I hear, "... running on a VAX 11/780." This is a typical airplane, I think — the natural habitat of electrical engineers. The person next to me is studying a computer manual. I have already hidden my technical journal. I don't want to talk technology on this flight; sometimes this stuff bores me, too.

Rattling around the back of my head is a disturbing image of something I saw at the airport. The back of my head is sort of a junk pile, but it's like a junk pile I've driven past and blocks later I get the feeling that there was something good being thrown out with all the trash. Now I'm remembering those giant piles of computer magazines right next to *People* and *Time* in the airport store. Does it bother anyone else that half the world is being told all of our hard-won secrets of computer technology? Remember how the lawyers cried foul when *How to Avoid Probate* was published? Are they taking no-fault insurance laws lying down? No way! But at the current rate it won't be long before there are stacks of the *Transactions on Information Theory* at the A&P checkout counters. Who's going to be impressed with us electrical engineers then? Are we, as the saying goes, giving away the store?

I've been reading that the U.S. government is concerned about giving away its technology to other countries. But shouldn't we engineers be worried more about all this technology going out to the public? Soon everyone will know how easy it is to design and program computers. How will we be able to keep our professional pride? I mean, you don't see any Heathkit home orthodontia kits, do you? You don't see any books on "How to Run Large Corporations from Your Basement in Your Spare Time," do you? Nope. But there are a number of companies that want every home to have a computer. Everyone will have the magical power, and then how will we be able to sell the assorted spells and incantations of our trade?

You may think that other professions could benefit from your expertise. Imagine, for example, talking with a group of brain surgeons. Shouldn't your electrical engineering background enable you to make a special contribution, they ask. With some hesitation you modestly suggest that maybe you could write some programs in BASIC on your Apple that would aid in the process of diagnosis. The doctors glance knowingly at each other, and then one smiles slightly and says to you that maybe this is a good idea you have had, but you would probably be better off using the battery of Cray I's in the next room and their specialized high-level language, Braincalc — or better yet, the array processors that have been designed for this purpose. Could you face such humiliation?

The time has come to put a stop to this drain of expertise. Bookstores should have to be licensed to dispense computer knowledge. Buyers should have to prove they are professional engineers in good standing with the IEEE. Computer journals should be kept behind the counter in plain brown wrappers, with the inscription "sold only for the prevention of computer crime." Certain advanced journals, such as *Byte*, should be sold by prescription only — such prescriptions being granted only when there is a severe "need to know." There should be special security clearance categories for the receiving of general engineering knowledge.

We should redo our educational system to protect our valuable knowledge. During the initial years of engineering school at college, the courses should consist of very basic mathematics and assorted humanities. Candidates should be carefully screened until their junior year, at which time the chosen few would be shown the secret handclasp, and the ritual of the dispensing of the Knowledge would be begun. Maybe we could erect a circle of large stones somewhere, and Fellows of the IEEE could go around wearing long robes. We could have lodges in cities around the world, with mysterious Thursday night meetings. No more of these shows like NCC where everybody sees what we're doing, and the press hovers about covering everything.

All papers submitted to the IEEE for publication should be carefully reviewed for sufficiency of turgidity. (Fortunately, not a great deal would have to be done here.) "Look here," the author would be told, "isn't there a more complicated way to say this? Can't you give a generalized integral equation representation over here, instead of this simple example?" Papers written in Russian and Japanese would, of course, be published in their original, untranslated formats as a special service to the membership.

IEEE should use its lobbying influence to pass laws forbidding the use of the binary number system by unlicensed practitioners. The use of BASIC would also be outlawed, and during an interim period until more esoteric languages could be evolved, uncommented assembly, LISP and APL would be the languages of the land. Very little would need to be changed in the present methods of documentation of programs. On the hardware side, the preferred embodiments of electrical circuitry should be small-scale TTL designs of processors, unless nonstandard, unmarked LSI chips could be obtained. A good deal of care would have to be taken in purging the literature of the mention of standards such as IEEE 488 and 696.

I know all this sounds difficult. Maybe it has all gone too far already, and the store already has been given away. But with concerted action perhaps it's not too late. With these and other similar measures in effect, perhaps the day will come when as you walk down the sidewalk children will be hushed and a path will form through the crowd. "It's an electrical engineer," one awestruck bystander whispers to no one in particular.

Part 2

So What if Engineers Can't Talk or Write?

Perhaps one of the subconscious reasons that I was attracted to engineering in my youth was its focus on things and problem-solving, rather than on people. I guess I had the idea that you could be like Edison or Bell, working by yourself in a cluttered lab on great inventions and philosophizing about the deep mysteries of life. Of course, it might be necessary occasionally to deal with an assistant ("Come here, Watson.") but on the whole engineering seemed a way of avoiding entanglement with the overwhelming social fabric. It was the polar opposite of, for example, being a car salesman.

As a junior in college I had to take a course in public speaking. I was more terrified of attending this course than just about anything else I have ever done. After sleepless nights I would tremble in the classroom, shrinking deeply into my seat in the hope of being unseen and unchosen in the random ordering of speakers to be called upon by the instructor. I kept asking myself why I had to suffer that torture. After all, I was to be an engineer, not a politician. I was so afraid of that class that I cannot recall a single thing that I learned there, only the feeling of the great ice age engulfing my mind whenever I walked into that well-remembered building. Even today when I hear that some friend has signed up for a toastmaster's club, where they give speeches just for practice, I feel incredulous, and I shiver with the conditioned response of fear.

When I finally finished that endless course on public speaking, I had the same feeling of relief that I got years later upon returning to earth (Newark Airport) after being a passenger on a 747 aircraft that had lost a section of its wing over Pennsylvania. You want to get on your knees and kiss the ground. Afterwards, the airline offered passengers alternative flights to San Francisco. "Not today, thanks," was the almost universal reply. No one even asked about frequent flyer miles for the aborted trip. I felt the same way about the speech course — not today, not ever again! I was so happy to be an engineer and not have to do this stuff.

Little did I know then! Sure, part of engineering is working with things — both abstract and physical. But no one will pay you just to do that. Not for long, anyway. Unfortunately, as an engineer you do have something in common with that car salesman. The difference is that instead of selling cars, you have to sell ideas, and in the process, yourself. Probably if they had told me that back in college, and in the unlikely event that I would have believed them, I would have looked for another profession. "Not for me, thanks," I would have said.

I remember the happy, early days of my industrial career, investigating theoretical problems in digital transmission. But the day soon came when I was scheduled to give a technical talk at an IEEE conference — the outcome of my thesis professor's insistence on publication of our material. Needless to say, I practiced being terrified for a week in advance. In my mind I began to plot my career so as to avoid any further necessity for such anguish and humiliation. "How can you go through with these talks?" I asked my wise and much-published research boss. I will never forget his answer. "You get used to it," he said.

I did not believe him, of course. No way. Advice like that was like whistling into a hurricane: not only wrong, but useless. Thus uncheered, I nervously approached my first podium as an engineer. Since that frightening experience I have stared from thousands of podiums into seas of blackness and blinding lights, and I have done my engineer thing. I often think of that early wisdom from my boss, and even as my very bones reject it, I know in my mind that it is true. You get used to it.

So I have gotten used to public speaking, and I have discovered a minor truth — it is actually not so hard. But in the process of that discovery I lost something very precious. I lost the fear but, more important, I lost what went with it — the euphoria that would accompany any successful, or even unsuccessful, completion of a public speech. I remember the lifting sense of freedom after a talk, the sheer exhilaration, the way I felt my feet and thoughts could take wing and, most of all, the sense of self-importance. But after too many talks it started to become just a job, something you do for a living. Now I think that perhaps a little fear is a good thing after all.

Boring Talks

As I made that long trip from fear to boredom I watched engineers as speakers. We're probably no better or worse than the doctors, lawyers, and business speakers I hear in their meetings, but we do have a special problem in the material we have to present. Basically, it usually comes out as boring. We have to face the fact that even we are mostly bored by it. (Though if you think this stuff is boring, try going to an insurance convention!) Now I happen to think that technology itself is pretty exciting, so that leaves only the presenters, namely us, as the purveyors of boredom.

I am continually intrigued at conferences by watching audience reactions. People are sleeping, or just at the brink of sleep. You can see in their eyes that no one is at home temporarily, like they have hung an "out to lunch" sign from their forehead. Other people are reading newspapers or skimming the conference program to see where else they might be. You get the feeling that they just ran into this session from another one where they were skimming the program to see where else they might be. They wander from session to session, moving targets ever in search of the "real" talks that they know must be there somewhere.

This boredom is the subject of the essay "The Gong Show," in which I describe a recurrent fantasy of mine in which speakers receive numerical feedback in real time from the audience. Not long ago I visited a university lecture room where I was surprised to see a keypad console on the arm of each student chair. In response to my question, my host explained that the instructor could get feedback on the quality of his talk by salting his lecture with questions requiring keypad responses. I thought to myself what a wonderful way this would be to gauge the ongoing success of your lecture, and I remembered my "gong show" column of years before. My host broke my reverie by commenting, "No one uses it."

I can understand why no one uses the feedback system voluntarily. That is the main theme of the essay, "Feedback." Praise is, of course, most welcome by everyone, but you and I both know that those impersonal buttons on the armrests are not going to give us praise; they are going to tell us things we would rather not hear. "Now I just want to check that you all are still out there," you say with a little nervous laugh. "Just press your button now so I'll know you're still alive." The scale tells you that 30% of the buttons have been pushed. Where do you go from there? Do you really need this?

I used to know a professor of psychology who conducted little tests with his students by choosing random times in his classes to interrupt his lecture and ask the students to write anonymously exactly what they had been thinking at that particular moment. No cheating, he would explain; they had to write what was on top of their mind. Probably you can get psychology students to do that; somehow I would not be so sanguine about using engineers and expecting honesty in anything so personal. "He has this handwriting analysis software," engineers would be whispering to each other.

The students' responses were correlated against the time into the lecture at which the interruption occurred. As I recall, some 75% of the students were thinking about the lecture material as deep into the lecture as 5 minutes. So far, so good, I suppose, but after only 15 minutes or so, he had lost the majority of the class. I think it was about at the 25-minute point that the results showed that the probability was about .7 that a given student was having a sexual fantasy. Perhaps again this is just psychology students. This would not apply to engineers.

Of course, the reception of your talk would be much improved if a certain segment of the audience were having fantasies rather than actively hating your talk. Unfortunately, I have proven time and again my rule of the 10 percent — that is, 10 percent of the audience will hate you and your talk. That is why feedback is not always positively motivating. "The speaker should have a deeper voice," says the feedback form. What are you supposed to do?

The other day I experienced yet another incentive to give up speaking. After my talk, the first questioner from the audience asked, "Are you aware of all the recent results in the field of cognitive disfribbling?" (or something like that). How are you supposed to answer a question like that? In the first place you cannot possibly be aware of all the latest results in the field of cognitive disfribbling. Furthermore, you are particularly hampered by not knowing what cognitive disfribbling is. But you can't admit from the podium that you don't know about some field that is obviously pertinent to your speech topic.

The unspoken import of a question like this is that, if you knew about cognitive disfribbling, you wouldn't be giving the talk you did. In other words, you are an idiot. Summoning up my disintegrating sense of self-worth, I answered, "Which particular results were you referring to?" (Of the many with which I am familiar, you pompous jerk.)

Banquet Speeches

I have a particular place in my chamber of horrors for banquet speeches. I confess to being amused myself at rereading "The Banquet Speech," in which I picture Sir Lancelot being the banquet speaker at a medieval engineering conference. I ask the unanswerable question of why we have to have banquet speeches. The eternal answer always comes back in the form of a question: What else is there?

I often find at conferences these days that old friends ask me out to share dinner on, say, Tuesday night. "That'll be great," I say. Then I pause for a second, reconsidering. "Isn't that the banquet evening ...?" I venture. For an instant they affect surprise, but they usually end with a knowledgeable shrug of the shoulders. Yes, that would be a good evening to find a restaurant somewhere else in the city. For us veterans.

Sir Lancelot represents for me the folly of the typical after-dinner speech. It is not just that it is boring and the hour is late. It is the hypocrisy of listening to the "famous" speaker when his speech has invariably been written by some anonymous speechwriter who would be so little esteemed in the engineering community that he would not even be welcomed at the conference. Of course, the famous personality may have inserted some of his own thoughts into the speech, or have ordered that they be included, but this is a hit-or-miss affair.

I remember meeting with a nervous speechwriter who had been ordered by a high executive to write a speech for a particular event. "Did he explain what he was looking for?" I asked. The writer complained that he had not even been allowed to meet with the busy executive. All he had been told was that "It had better be better than last year's talk." I would have been nervous, too.

I used to have this fantasy of publicly abdicating a talk when the actual occasion arose and it did not seem appropriate. As it turns out, I actually did this once. I no longer have that fantasy. I had agreed to give the after-dinner talk at an IEEE awards banquet. As you might imagine, the proceedings droned on forever. It seemed that there were awards for everything, and that if we waited long enough, everyone at the banquet would be given an award. And then it seemed that every recipient had been watching too many Academy Award shows and had come up with some interminable thank-you speech, thanking everyone down to and including their dog.

By the time I was introduced for the after-dinner talk, it was after eleven o'clock and both the audience and I were weary with pomp and drink. This is a bad idea, I said to myself. If ever there were a time to put my fantasy to the test, this was it. I started my talk by telling the audience that in these circumstances a talk was a bad idea, and that I would go through with it for only a brief period. I further explained that I had preselected a minimum time, and had set the alarm on my watch. I wouldn't tell them how long else they might time the talk themselves. When the alarm sounded, I would sit down without saying another word.

All this explanation took some saying, and just as I finished stating the ground rules, the alarm went off. I sat down. In my mind I imagined tumultuous applause for my having the courage to abort a talk that no one wanted to hear. In fact, people were stunned, mystified, and perhaps embarrassed. There was very little applause. I suppose they pondered the same dilemma I face in considering whether to leave a tip at a self-service restaurant.

A week or so later I was in the back of an elevator, and I overheard someone talking about my nonspeech to some other people in the front. "Can you imagine?" they were saying indignantly, "He didn't give his speech." I heard the audible sighs of shock, like I had committed this unspeakable crime, to make a pun.

I had much occasion later to think about this nonspeech. I will never be convinced that the people actually present at that moment wanted to hear a speech from me, or from the President, or from anyone else. Yet there was a social compulsion. We will go through with this, it demands. It is thus written, and thus it must be. It is a torture in which we must all meekly play our roles. I consider how I feel when I see someone in the audience walking out of a talk. I feel jealous and angry. How can they get out and not share this misery with the rest of us? Strangely, perhaps the speaker is not viewed as the creator of the misery. That role is served by whoever convinced us to go to the banquet in the first place. The speaker must line up and share the misery with everyone else. We are all in it together.

Engineers and Vugraphs

For most engineers the thought of giving an after-dinner speech, as opposed to just suffering through the listening end of one, is especially scary. The reason is that you will not be able to use vugraphs. What would we do without our vugraphs? We have come to rely completely upon a pile of transparent foils, as if they were giving the talk and we were just standing uselessly beside the overhead projector. Take away our vugraphs and what do you have? A speechless engineer.

In "Sliding By" I used the theme of the ubiquity of vugraphs in the life of an engineer. I do not know whether we are more visually oriented than other occupations, but in my company there is an oft-spoken jest that no engineer is capable of giving a talk without vugraphs. It is the kind of nervous half-jest that is used by an engineer to explain the large stack of vugraphs he has brought to an informal meeting in which there is some small chance that he might have to say something.

We also have this propensity to take far too many vugraphs to any speaking opportunity. This is the subject of "Not Ready for Prime Time." I do this myself almost all the time, thinking erroneously that I may have to delve more deeply into this or that subject and imagining that I must be prepared to take this or that turn in my talk. As a result of this compulsion to think of a simple talk as some kind of hypermedia event, the talk itself loses its thread, and time expires while many key vugraphs remain unshown. Then I do the same thing everyone else does: I talk faster. What else is there? I could not possibly — not ever — skip a precious vugraph.

Better Technical Talks?

I am sure that talks by engineers will be good material for future columns by me and others. We are certainly easy to criticize, but that begs the question: How do you give a good talk? Unfortunately, I am not so sure about that. I only know what seems wrong. What seems most wrong is that we so often fail to communicate the simple essence of our work — what did you do, and what results did you get? Above that, what I look and listen for is some personal added value from the speaker in the form of little wisdoms that they gained from doing whatever it was. This is precious stuff.

What gets in the way of this simple-sounding formula is our pretensions. We just take ourselves too seriously. A case in point is our dull, formal, and sometimes pompous biographies. This is the subject of "Credentials: Who Needs Them?" We hide our real selves behind these engineering credentials that all sound the same. You got born somewhere or other, went to such-and-such school, got a job — that's all there is. I have this frustration about wanting us to be more human and more individual, but I too feel trapped in the rigid convention of our profession.

After the essay on credentials was published, someone sent me a copy of a biography that had been published in the *Proceedings of the IEEE* in 1979. It said that the author had received the Ph.D. degree in physics "after exhaustive study of an esoteric topic of no practical value." After that he "outraced scientific obsolescence" through promotion, and was currently in pursuit of the Peter Principle (promotion above the level of competence). He boasted "the usual affiliations that carry arcane Greek letter designations in return for the payment of annual dues, as well as membership in technical societies that offer cheap life insurance." Unfortunately someone else sent me a biography that he had written for an IEEE journal that told of some of his hobbies and outside accomplishments. It had been rejected as insufficiently professional.

The mystique of engineering weighs heavily upon us in giving technical talks. No glimmer of incompetence must be seen. All of us are afraid that someone in the audience knows more than we do about whatever it is, and all of us are out to impress our peers. So we show our mathematics, make the problems seem very hard and mysterious, and talk about very general cases. Yet when we are on the other side of the podium, as listeners, this is exactly what we do not want to hear. The people in the audience do not care about our math or our clever little tricks. Strange how our values change when we make that little walk from podium to audience.

An example of our pretensions getting in the way of simple expression is the stilted vocabulary developed and used by engineers. We talk differently on the stage than we do off, and usually not for the better. I often have had the experience of seeing a friend get behind a podium and asking myself, who is this person? Did he accidentally get caught in a full moon or what? Why is he talking that way? Something that he would have explained to me in simple English a few minutes ago is being distorted beyond recognition in a groping search for what he believes will be proper, formal expression.

Then, to make matters worse, we employ our own secret engineering language, in which it is not allowed to use terms like "thingamajig" and "widget." Everything has to have a special name, so that outsiders will not know what we are talking about. Unfortunately, we ourselves are often the outsiders when we stretch our interests the least bit. Did you ever wander into the wrong session at a conference and not have the least idea what the conversation was about?

Of course, every profession and occupation develops their own special vocabulary, and I do not think we engineers are the worst offenders. I remember well my first military meeting at the Pentagon. Every second word was an acronym that I had never heard before. I would keep trying different interpretations of each acronym in my mind, like trying to solve some crossword puzzle. The curious thing about this exercise is how it used up my mind, and I would miss all the subsequent words that the speaker was saying. Then I would be three or four acronyms behind on my push-down stack of unknowns needing to be interpreted. Pretty soon my mind turned to mush, but maybe that had happened long before anyway.

In spite of our own difficulties when other people use acronyms, the temptation to start throwing your weight around with your own acronyms is overwhelming, like you are in the know, too. I can't help it myself; it makes me feel so knowledgeable.

Aside from acronyms and technical names, another facet of our conversation is the overuse of certain fashionable "power" words. That is the subject of "A Way with Words," which reflects on the strange way certain words leap from centuries of obscurity into prominence. If you are "in the know," you will use these important words liberally. Examples of recent power words are *process* (as in *business process*), *platform,* and *object-oriented.* Use them in good health — my gift.

One characteristic that these special words have is that no one is quite sure what they mean, but everyone understands implicitly that, whatever their meanings are, they are crucial to what is being discussed. Thus, the use of these fashionable words will deepen the significance of your talk.

The problem is to project an easy, confident expertise without actually going into details or difficult material. This is especially hard when you are just starting your career. No one knows you, and there is a palpable sense of reserved judgment in the audience. "Go ahead; try to impress us," they seem to be saying. Furthermore, all you really have are the details. It is hard to give the big picture when no one has let you step back and see the vista from a good vantage point. But you have to try, and people will appreciate that. How does what you did fit into the larger context? What did you do that was different? Forget all the math and the crank-turning stuff.

Unfortunately, like a lot of advice this is easy to say and hard to do. In golf I always used to tell myself not to try to kill the ball. "Don't kill the ball!" I would shout in my mind. Then I would step up to the first tee, and everyone would be watching. The pressure would be on. Everybody would be really quiet so as not to disturb my swing, but of course that intense quiet would only serve to increase the pressure. I think the onlookers know this, and they concentrate their minds in a form of voodoo in order to mess up your shot, so they will not look so bad when they mess up themselves. It is usually very effective.

There are some parallels in giving talks. You are much better off in speaking if you don't try to kill the ball there either, and by this I mean trying too hard to impress people, particularly by trying to say too much and going into too many details. Just keep it simple. The difference between golf and speaking is that even though you may feel that the quiet from the audience for your speech is ominous, they really do wish you well. They want you to succeed almost as much as you do. If you can latch onto this psychic wave, you can ride it to success.

Of course, not all audiences are receptive. I often think of a talk as a dialog, and the audience must play its part effectively. You can listen to what the audience is saying to you during your talk — by their expressions and their body language. In a good talk some people will appear to be attentive and rapt, but it is inevitable that others will have turned you off, and some will even be sleeping. Try to tell yourself that it is not your fault. This is a consequence of my general rule of the unavoidable 10% who dislike you and your talk. I always tell myself it is not my fault. I never believe it, though.

Some audiences are better than others. I have often given an entirely similar talk to two different audiences, and have felt successful with one, while feeling like a failure with the other. It is not just the particular people in the audience that make the difference, but even the physical setting. The worst situation is either at the end of the day or right after lunch, in a concrete auditorium that is too large, and with attendees sparsely distributed. The best setting is a comfortable but crowded ballroom first thing in the morning. But most of us don't get the latter setting; we get the former one.

There are lots of books and pamphlets about how to give a good technical talk. I cannot believe that some of them advocate writing out your talk and then reading it to the audience. Please, never do this. It is guaranteed to be irredeemably boring.

Something that almost all advice booklets recommend is rehearsal. This is the kind of thing that is good for other people, but not for yourself. When you give other people advice about their talks, you feel very magnanimous and wise. How nice of you to have helped them so much. But when you are the person being helped, it is an entirely different story. Everyone gives you conflicting advice, and heaven help you if you have more than one boss giving advice. You end up not owning the talk, and your worst fear is no longer giving the talk itself, but messing up the careful wording that management has advised in crucial parts of your talk.

Rehearsal also can take the life out of a talk and leave it juiceless. Enthusiasm is the most important ingredient of any talk, and it can easily take wing and leave you flat. When I did a television interview with Bill Moyers, he refused to have lunch with me before the filming. The first time I actually met him we were on camera, and as a result the interview had a certain spontaneity — for better or for worse. On the other hand, when I did an interview with Jane Pauley for the Today Show, we did have lunch together, and talked about things that might be covered. Later when they turned the cameras on her, she underwent such an instantaneous metamorphosis into her television persona that I could not handle my end. The camera beamed its red eye at me. "Cut!" I said. Things do not always work out the way you envision. Nevertheless, I must confess that I have seen talks that were improved by rehearsals, and talks that should have been improved by rehearsals. I just don't want to go through them myself.

Technical Papers

I am surprised that in ten years of writing "Reflections" I have written only one full column, "The Papermill," about our technical publications. They deserve a lot more analysis. My problem is that I cannot figure how the system works. The whole business of technical papers has been bequeathed to us from antiquity, and we seem to be stuck with it.

The office in which I am writing these sentences is walled with shelves of technical journals. They look important. When visitors come to my office they always look impressed. I feel good in the presence of all this knowledge, but there is a dark secret here: I don't read them. Please don't tell anyone.

This is such a high-level secret that I carefully keep it from myself — in my heart I believe that I read these journals. In any event, I absolutely know that I *could* read these journals. If I wanted to. It is just that I have been too busy. Besides, I know what is in them. Everybody seems to know what is in them, more or less. The mystery is who reads them in the first place and tells us all.

The world works in strange, unfathomable ways. Most papers are probably read by almost no one — maybe not even by the reviewer. But when a good paper comes along with the germ of a new idea, somehow the word gets out on the street. A few people really read the paper, and perhaps it gets translated from mathematicalese to English. Eventually it becomes part of the common lore. Maybe it even turns into a product.

The point of being an engineer is the thought that you could read those technical papers if you wanted to. Years ago I was in a group of people in a lunchtime discussion, and the subject of a particular recent technical paper came up. One of my friends commented that he hadn't read that paper, but of course, could if he wanted to. This time his bluff was called, though, and bets were taken.

I remember the erstwhile reader sharpening a number of pencils in order to take notes while he read. About five of us stood waiting expectantly outside his closed office door, watching the dial of a stopwatch. How long would he last? No contest, really. After only about five minutes he stormed out of the office in disgust, throwing down one of the still-sharp pencils. "I can't read this stupid thing," he muttered. For myself, I think it is better never to have your bluff called.

There is a problem with technical papers, as compared with talks. Talks evaporate into the air as they are spoken, but papers hang around indefinitely. In our technical world things have a habit of becoming quickly obsolete. That paper you were so proud of not long ago becomes an embarrassment. I have heard of people planning to steal into the library late at night for the purpose of cutting their papers out of journals. They really should not worry — no one is going to read them anyway!

The Gong Show

I'm writing this column while a dull meeting washes about me. Of course the words "dull meeting" are redundant. It is the curse of our profession. A speaker is holding forth on a topic of no apparent interest to anyone in attendance. The person beside me is asleep, or has perhaps expired. Most of the audience is composed of Stepford engineers, whose unfocused gazes fail to penetrate the fog rising through the cone of projected light. The Vugraph machine emits its patented hypnotic hum, rendering sterile all persons within a 4-meter radius. (Attention Vugraph's lawyers: I'm only kidding!) Do they have Vugraphs in China, I wonder, or is this a uniquely Western form of indoctrination?

The speaker doesn't want to give this talk. I deduce this from the violent trembling of his entire body. The audience doesn't want to listen to this talk. I glean this bit of information from the mass stupefaction I observe. So why are we doing this? Is it some sort of ritualistic test of engineerhood? Why doesn't someone jump up, wave his arms, and shout "Cut! You can all go home now"? In my fantasies I see myself actually committing this piece of treachery. What would the people here do? What would you do if someone did this in a meeting? The urge is so overpowering I may actually do it any minute now.

I've been involved in setting up many meetings — IEEE and otherwise. Somehow when you're on the program committee it all looks so good on paper. Joseph Buffoon, the noted expert on whatever it is, has agreed to give a keynote address on whatever it is. His credentials look great in the advance program and all the flyers being mailed out. Unfortunately, those in the know know that good old Joe is the world's worst speaker — a title for which there is much competition. On the day of the actual talk, such as it is, there is a great deal of groaning amongst the audience, and the program committee fields the complaints with a sense of betrayal. Joe has failed to give the talk the committee had in mind. However, the committee is now dissolved, and with it the small amount of corporate memory gained. Next year's committee feels that Joe would be ideal to give a principal address on a new subject that is now important to those who will attend, and we iterate from there.

I have a secret plan for better conferences. On the armrest of every seat in the auditorium is a button. Behind the speaker, but out of his view, is a large digital display. At any given time the display reads the percentage of the audience who have "pushed the button." The buttons, needless to say, are to be pushed when one no longer wants to hear any more. The tension mounts as the percentage figure behind the speaker creeps closer to the agreed-upon cutoff figure. Perhaps the speaker senses the impending mutiny of the audience and begins speaking faster and faster, hoping to get in as much of his talk as possible before his demise. Of course this would only accelerate the inevitable climb of the mute digits. There would be some sort of ceremony on reaching cutoff. Perhaps the public-address amplifier would automatically switch off. In my dream I see a giant hook descending on the unfortunate occupant of the podium, but that might really be going too far. After all, it could well be me on the end of the hook.

There have been many attempts to improve the quality of talks at conferences. Rehearsals have been held, organizers have worked with speakers, and booklets of advice have been prepared and mailed. I remember one booklet with explicit good advice. "Hold your slide at arm's length," it said. "Read the caption. If you cannot read the caption, open fingers, releasing slide." Another ploy that has been used is to give an award for the best presentation at the conference. Doubts about the effectiveness of such an inducement led the awards chairperson of the IEEE Communications Society to suggest that an award be given instead for the *worst* talk at the conference. This award would be given with much fanfare and would be suitably publicized in IEEE literature. When this suggestion was made at a Board of Governors meeting, a thoughtful hush fell over the long table. "Yes, that could work," each of us thought quietly, "but would we dare?" There was no discussion of the suggestion.

Many years ago I attended my first conference at a small, out-of-the-way place. It is memorable for two reasons. First, at the banquet I inadvertently sat at the award-winners' table. When the persons at my table were given awards for various reasons, one by one around the table, I was at first filled with embarrassment. But, sensing the inevitability of it all, I became convinced that I would receive some sort of award when my time came. I didn't, but that's another story. The other memorable event in that conference was the worst presentation I have ever heard. Naturally, that covers a lot of ground, and it's only with a sense of awe that I recall this particular talk.

It was an after-lunch session in one of those sloping concrete rooms that radiate indifference to humanity. There was no light dimmer, so all the lights were out, and the room seemed lit only by reflections off the myriad dust particles in the slide projector's beam. The speaker read his talk verbatim without ever looking up. Either he had never learned to read properly or, giving him the benefit of the doubt, he could not see his manuscript in the dim light. In any event, he effected a combination of inaudibility and incomprehensibility. There seemed an infinity of slides as he droned on and on, pointing out what he felt were salient characteristics of various curves. None of it, and I mean none of it, made any sense. One of those moments of great clarity came to me when I felt I was the only person alive in the room. "We aren't really, actually going through this charade," I thought. Finally the speaker said, "Last slide, please." After a long pause, the voice of the projectionist came from the rear of the room. "The other one was the last slide," he said. "Oh dear," the speaker apologized, "I've been out of synchronization. Perhaps I'd better start over." The session chairman leaped forward like the Lone Ranger. "That won't be necessary," he said quietly, but emphatically.

During the next talk I was reflecting on the previous talk as an art form, perfect in its kitschlike qualities. My reveries were shattered by a loud crash, followed by ripples of satellite crashes down the center aisle. When the lights were turned on, the broken projector lay in a heap tangled with the infamous previous speaker. "I just wanted to get my slides," he mumbled to the stunned audience.

I guess it's too much to ask that our conferences would be full of scintillating talks on fascinating subjects. Is it just engineering? Is it us? Or is it just human? If you have a plan, let me know.

Feedback

I was about to realize a fantasy. On the scale of fantasies, this was not one of major proportions, but life offers few enough such opportunities. Glancing once again at the stranger sitting next to me during the long plane flight, I monitored the progress of his page-by-page study of *Business Week* magazine. Lying on his briefcase at his feet were two other magazines presumably awaiting the same thorough scrutiny. The one on top was *IEEE Spectrum* — an issue containing one of my own columns.

I have never actually seen any real person reading something that I have written, and so I nervously anticipated his reaction. What better feedback than to gauge the rapport emanating from a total stranger cramped up against you as he reads words that not long ago had only been images in your brain! Even though in everyday conversation we see the effect our words have on the listener, the medium of writing is another matter entirely. The writer and the reader are separated by a chasm of distance, time, and anonymity. The loop is not meant to be closed.

As my seatmate reached for *Spectrum*, I took the opportunity to study him a little more closely. He would be a demanding critic, I thought. For one thing, he wore a vest, so I leaped to the conclusion that he was a formal sort of person. Moreover, he looked ... well ... tough. Nonetheless, I was ready for the challenge.

Finally he reached the penultimate page, and I tuned my sensory perception to its maximum level. As the page was turned to my column, he folded the magazine back to get a better reading platform and settled into his seat. But before he had even gotten comfortable, he emitted a grunt of exasperation and turned the page decisively. After giving a kind of snort of derision and a little shake of his head, he resumed his in-depth reading of the material following my column. I had been ripped off — just like that!

Ah, well, such is life. Somebody once said that you cannot please all of the people all of the time. Intellectually, I know that, but it is a difficult principle to apply. Some years ago, I had to give a series of talks in a management forum. In order to help with the quality of presentation (so they said), the organizers collected feedback on each speaker. Shortly after each talk, I would be given the compilation of grades and comments on my presentation. I can assure you that was infinitely depressing. I discovered for the first time what I consider to be a fundamental law of public speaking: 10 percent of the people will hate you.

I remember one talk in the series especially well. My subject was "technology" — as opposed to "administration," "leadership," "business," and so on. As soon as I was introduced, a member of the audience raised his hand to ask a question. I had not yet even said a word. "Can you define 'technology'?" he asked. I stammered a bit. I mean, we all know what "technology" is, don't we? When I got the audience's feedback from that abortive talk, one of the forms rated me in the lowest possible category. "Speaker could not define 'technology' " was the added comment.

I got so gun-shy that I would look out at the audience before my talks and wonder who were those 10 percent out there who were going to dislike everything I said. Maybe they were the same people each time, being moved from talk to talk. As I was speaking, I would examine the audience for little indicators — such as falling asleep — that betray these secret antagonists.

In the years since this unfortunate revelation, I have had many subsequent occasions to verify the principle of the unimpressed. Now when they offer me feedback on a talk, I politely decline. I would rather imagine that the audience liked me than know the truth. Besides which, I have developed yet another theory about audience feedback: not only is it demotivating, but it is even unhelpful. One person will write that the talk was too theoretical, while another will say that it was too practical. One will say there was too much introduction, while another will say too little. The best you can hope for is to balance all of the negatives, but this hardly leaves you feeling very good about the whole thing. Balancing positives would leave a lot better taste.

I believe that most people have an intuitive sense about how well they themselves have done on a particular speech, writing, or job assignment. Deep down, we know the truth, so most of the time we prefer to hear lies or platitudes. In the case of a speech, I often feel that it is a real-time event that is a collaboration of speaker and audience, so I rationalize that the audience shares my ineptness. Sometimes I give a nearly identical speech to two different audiences. One audience seems to think it is great, while for the other it falls completely flat.

Even when people praise your work, you are suspicious — sometimes with good reason. Not too long ago, I gave an after-dinner talk at some university. I did not do a good job. Afterwards, the departing streams of people formed two separate columns about me, creating a little lonely vacuum in the vortex that I alone inhabited. One elderly man approached me. "I really enjoyed your talk," he bubbled. I brightened. "If only the batteries in my hearing aid weren't so run down, I would have gotten more of it," he added. I dimmed.

Recently I participated in a meeting about something or other. It was probably terribly important at the time, but I have forgotten. What I remember was that afterwards a woman whom I had never met came up to me. "You're really a very funny man," she said. I wasn't sure what the correct response to this statement was, so I mumbled something incomprehensible and shook my head diagonally. She turned to leave, but as a sort of afterthought, added, "I don't know if you mean to be ..." She left with the unfinished thought hanging in the air.

The other day I got feedback that for some perverse reason has bothered me unreasonably. About one of my talks, someone in the audience wrote, "OK — but not up to his usual standard." I thought, what kind of criterion is this? Do we always have to be better than we were on previous occasions? Are we never entitled to off days? Have mercy!

On the off chance that you are reading this on an airplane, smile at your neighbor. You never know.

The Banquet Speech

Why do we have after-dinner speeches? Has anyone ever asked? As long as I can remember, this cruel and unusual punishment has been inflicted on conference-goers. All day long at the meeting you spend endless hours in darkened rooms listening to talk after talk. Finally it is time for cocktails and a mass-produced, synthethic dinner of thinly sliced, graying mystery meat accompanied by 27 peas. A thousand people are interminably served as you discuss the weather with sullen strangers cramped knee to knee in the the smoky heat. Long after the coffee is cold and the digestion process has drained the blood from your brain, through the glazed fog you see the emcee take the podium. "And now what we have all been waiting for ...," he begins. And what is it that we have been waiting for? Another talk. But not just *any* talk; what you soon discover is that this talk is the most boring of the entire conference, thus deserving its special showcase. Why do we have after-dinner talks? Has anyone ever asked?

Was it always thus? I imagine how it must have been for medieval engineers. The conference announcement is posted in all the town squares. In the proclamation even the subject of the conference — having to do with the relative advantages of storming a castle by pummeling a door versus breaching a wall — is given little attention relative to the glowing portrayal of the banquet speaker, Sir Lancelot, who will speak on "The Joy in Purity of Heart and Avoidance of Unspecified Sin." This must truly be an important meeting, the knight engineers tell their feudal lords, for look who is speaking at the banquet! Because of Sir Lancelot's inclusion in the program, the necessary military service waivers are obtained from various feudal armies, and attendees pour into the castle from throughout the land.

All day long at the symposium the knights clank in and out of the dank meeting rooms, carrying large stacks of overhead oilskin transparencies in their heavy gauntlets. Voices drone on in the hastily converted dungeon facilities, punctuated by semi-intelligent grunts and interrupted occasionally by the thunk of a falling visor as a helmeted head nods. Finally it is time for the great banquet. Here I imagine a scene from *Tom Jones* or *Excalibur,* with long, rough-hewn wooden trestle tables laden with wine and platters of food. Through the tumultuous din the servers pass among the knights collecting the banqueting tickets — red for hog, and blue for the low-cholesterol gruel.

The malodorous fires in the giant fireplaces have died to embers, and more than a few armored knights lie corroding in pools of wine by the time Sir Lancelot mounts the podium for his after-dinner speech. Not so much a hush, as a restructuring of the cacophony, passes over the audience with his appearance. The great man is there! Sir Lancelot ponderously clears his throat, and begins laboriously reading from an illuminated manuscript contained in a black, sheepskin valise.

Sir Lancelot speaks! But what is he saying? Presumably this is English, as that is the official language of the conference. The sentences seem to have grammatical structure, but no resulting information sticks as a residue in the listeners' minds. Among the rapt audience only the neglected court jester and a large shaggy dog, damp with wine, seem to realize the truth. Sir Lancelot is a stiff.

What makes the situation even more ridiculous is that the words are not really those of Sir Lancelot at all, having been written by an apprentice speech-writer whose station in life is far below even the squire engineers in the back of the hall. Lancelot's only contribution has been to order that no information, opinion, or value judgement shall be contained in the speech. This is partly because he fears being quoted by the town criers, shouting the headline, "Spokesman for King Arthur says ...!" The danger that Sir Lancelot will embroider upon the written speech is negligible; it has been so long since he actually razed a castle or succumbed to unspecified sin that he really has no opinion of his own on the subjects anyway.

Still later, when the forest of grease candles is but a field of stumps, only the dog looks up in puzzlement at the sudden quiet as Sir Lancelot, finding no more pages in his valise, abruptly ends his speech at a seemingly premature point. No one cares. There is polite applause from a fearful few. Why do we have after-dinner speeches? Has anyone ever asked?

Of course the problem that every conference organizer faces in considering the banquet speech is — what else is there? I have a proposal. My idea had its genesis in the memory of a pair of IEEE conferences that I attended some years ago. At the first, a systems conference in Miami, U.S. Vice President Hubert Humphrey was the banquet speaker, while at a communications conference in Los Angeles the former Israeli Minister of Defense, Moshe Dayan, filled that role. In each case the conference organizer was concerned about the proper treatment of his illustrious visitor. My advice to each was the same. "Don't worry," I said. "They won't show up." Neither did. The conferences were wonderful; no one minded at all.

My plan is this; a most famous person (MFP) is invited to be virtual banquet speaker. His or her picture appears prominently in the brochures and programs, lending prestige to the conference and authenticating its significance. Then, during the actual conference, the MFP is obligated to attend one entire session, selected at random. The MFP must actually sit among the conferees and pretend to be interested. The evening of the banquet, the MFP is a guest at the public cocktail party, where he or she is required to mill about among the guests. If the conference chairperson wishes to have a private cocktail party in his hotel suite, as is so often the custom, he may. The MFP will be invited but told he need not be there. So he will not be. Finally, when the dinner finishes, the program concludes with — nothing! Freedom!

Wish me luck with this daring proposal. The sanity I save may be your own.

Sliding By

Does anyone remember what we did before there were photographic slides? I thought not. The beginnings of our slide-oriented culture are lost in antiquity. They go back to a time when the universe was young and the earth a forgotten planet in a faraway galaxy. Long before there was life as we know it, out of the primordial slime the early forms of slide transparencies began to congeal. Early humans puzzled over the meaning of these strange, thin transparent slabs that littered the primitive landscape. The use for which these transparencies were destined awaited the discovery of fire — the first source of illumination. It is not known exactly how or when cave dwellers began to use slide presentations on the walls of their caverns to depict plans for animal hunts, but the remains of some of these images can still be found. The history of transparencies projected overhead is likewise clouded in obscurity. Nevertheless, what we have learned over the years about slides applies almost universally to overhead transparencies, too.

For many thousands of years the only type of slide was pictorial. Slide presentations were generally simple and easy to follow. ("You strike *here* with the club.") Eventually primitive people made an accidental but important discovery — a slide transparency with a series of heavy black dots in a vertical line. Perhaps a blank transparency had fallen into a bed of ashes, but in any event the "bullet" was born as a graphic element. Seeing such bullets, people realized that an important format had been found but that something was missing. Thus writing had to be invented to fill in behind the heavy dots. Slide presentations were never the same after this.

The role of slide presentations in history has often been overlooked. For example, one of the great briefings of all time was the all-day slide lecture given by Columbus to Queen Isabella in his request for funding an important exploratory project. Fletcher Christian went through hundreds of slides in a series of presentations in the audiovisual room of the *H.M.S. Bounty* to explain a proposed management reorganization. In science and technology many great inventions were realized in celluloid format. Thomas Edison displayed slides of his proposed light bulb on an early model projection machine that used gaslight. Naturally, the incentive for his great invention was improved audiovisual presentations.

In today's sophisticated contemporary engineering environment slide presentations are the stuff of life. The average manager must be up to consuming hundreds or even thousands of individual slide views per week. This requires years of practice and daily exercise. Some high-level executives recommend a strenuous program of early morning warmups with a home projector, to be fit for the day's activity.

On the other hand, a different kind of stress takes its toll on the employee who must make the slide presentation. Of course there is the usual fear of public speaking, but at the back of this employee's mind lurks the fear that something will go wrong with the slides themselves. Consider for example the feeling you get watching the well-known television commercial about an overnight delivery service in which the would-be presenter makes gestures representing animal silhouettes in front of an incredulous audience, because his slides didn't arrive in time. Hits home, doesn't it?

Success in engineering today depends on knowing how to prepare and use slides. Most graduating students have had little or no experience with this. Thus I thought it would be a service to the profession to include in this column a number of pointers for the neophytes.

1. Take all the slides you own to the presentation meeting. If you are new, or have just arrived on earth and don't own many, then borrow some. The stack of slide trays should be very high. This puts the audience on edge, and establishes you as an expert in whatever it is. You may begin by saying that you have brought more slides of charts than you need, but start from the top of the stack. The audience will have no idea when, if ever, you intend to finish. This helps discourage questions. If you are especially worried about questions, take several trays of back-up slides. After you have dipped into this pile to handle the first question, there will be few further inquiries.

2. It is wise to begin with a series of "wiring diagrams" — charts showing the organization of your institution. Never underestimate the interest of your listeners in the myriad details of your organization.

3. If you prefer overhead transparencies you can exploit a special technique: never use several transparencies when the same effect can be accomplished with a single transparency and a complicated series of fold-down overlays attached to it. Your expertise in handling this advanced format will be noted by management.

4. Most of the charts you show should be of the "bullet" variety, with no less than 12 major bulleted points per chart. This makes your timing easier to calculate. Allow 30 seconds per bullet.

5. No real engineer places slides upside down in the projector. (We went through all those intelligence tests on visualizing rotations and permutations.) It is, however, good form to occasionally cock a slide projector at a 20-degree angle. Pay no attention to the leaning of the audience. This keeps them slightly off balance and prevents sleepiness. This maneuver is even easier using an overhead projector. The angle can then be casually varied from transparency to transparency, thus defeating those members of the audience who have learned to doze at a fixed angle.

6. Incorporate a few slides showing giant tables of small print. Always apologize for these, saying you know they can't be read, but ... (Save them carefully for subsequent talks.)

7. A sense of timing is critical to successful presentations. Watch your audience very carefully during each chart for telltale signs — such as a slight widening of the eyes — that might indicate the dawn of understanding. The slide should be then be removed immediately. After a few presentations with a given set of slides you should get some feel for the timing on each chart. This is known as the mean time to failure of that slide. The total presentation time is simply the sum of the mean times to failure of the individual charts.

This last point is courtesy of General Glenn Kent (and he is real — Lt. Gen. USAF, ret.), whom many readers will recognize as one of the great briefing-debunkers of all time. "This is just like a good old-fashioned American game of baseball," he is fond of saying. "There are pitchers (briefers) and batters (briefees)." Gen. Kent counsels young briefers (usually, he notes, pale captains delivering briefings on behalf of suntanned colonels) to practice their fast balls, curves, and change-ups. There is always the ultimate fear awaiting — the heavy hitter. He could be the one person in the audience who understands something you have shown sufficiently well to ask an embarrassing, unanswerable question. A carefully prepared presentation, however, can cause even the mighty Casey to strike out.

Not Ready for Prime Time

I was to be the last speaker on the afternoon's agenda of business briefings. Each of us had been allotted 20 minutes to make our pitch. I glanced sidelong at the other speakers. Like the supermarket customer waiting in the express checkout line — 10 items or less — I began to count the transparencies that each speaker clutched in his lap. I lost count, discouraged, somewhere in the mid-20s on the first speaker. Each of them had far more than I did — and I had too many myself.

Is it that we are really incapable of estimating how much time each briefing chart will consume? That started as a rhetorical question, but I am afraid that the answer is "yes." Even down deep, we fool ourselves; after all, the motivations are all wrong. The actual time consumed by a given chart will be determined by statistical imponderables beyond our control, or so we think. In the face of these cumulative uncertainties, we aim for that distant goal 20 minutes down a long, winding road whose road signs are our precious charts.

Granted, we cannot hit the endpoint exactly in time, but shall we raise our sights and aim long, or shall we lower them to an imaginary shorter point? Not much of a question really. If our talk runs long, probably no one will notice. At worst some chairperson will ask us to finish in a few minutes. But what happens if we finish ahead of schedule? That scenario is so horrible to contemplate that we thrust it from our minds — and shudderingly add a few more transparencies to our pile, just in case.

Imagine finishing your 20-minute talk in, say, 12 minutes. The audience of your assembled bosses looks at you expectantly, waiting for you to continue from this inappropriate pause. "That's all I have to say," you mumble semiaudibly. One boss glances at another, raising his eyebrows minutely. You sense that it is the "that's all he has to say?" raised-eyebrow expression. Another boss uncrosses her knees and jots something on a notepad. Obviously, she is writing something to the effect that you had nothing more to say, having exhausted your entire knowledge of this subject in a mere 12 minutes.

I wonder if anyone has ever compiled data on the actual length of short presentations as a percent of the allotted time? There are probably no known instances of anyone actually finishing early, so the histogram starts at something over one hundred percent and tails off slowly past two hundred and three hundred percent, and so forth.

Of course, it is very hard to give a short talk. A longer talk is much easier. For an hour talk, little planning is required; you simply tell everything you know about whatever it is. On the other hand, a 20-minute talk requires organization and forethought. Mostly it requires throwing out some of the precious material that will show the audience how smart you are. Oh, that is hard! A 10-minute talk is even more difficult. Generally speaking, talks get tougher to give as they get shorter and shorter. But there is something like Zeno's paradox here, because finally a talk of zero length is very easy to give!

Thus at some non-zero length, a talk becomes maximally hard to give. Ordinarily, I would advocate research to identify this magic interval, but it has already been uncovered by the media people — it is the elusive sound bite. If you ever have the experience of being interviewed by the media, that is what they will be looking for. In order to make the 11 o'clock news, you have to say something devastatingly important in about three words. When your big opportunity arises, the interviewer will ask you some ambiguous, long-winded question. The red, glowing eye of the television camera looks at you, and you hear the silence roaring at you like Niagara Falls.

"There are many considerations associated with that question," you begin. The expectant smile on the interviewer fades. "But first, I have to assert some qualifications," you continue. The interviewer glances back at the camera person. A look passes between them, and the camera person gives an infinitesimal shrug. He looks away and rolls his eyes in subtle disgust at your ineptitude. Meanwhile, you have lost your train of thought ....

But back to my present predicament. The first speaker has now used 5 minutes on the first chart alone. I keep willing him to change charts, but to no avail. I know how he feels. When you first start a talk, it seems that you have forever to fill. Time expands, and the audience hangs on your every word. Your arguments sound amazingly convincing, and you conceive of little explanatory digressions on the spur of the moment. That is, of course, at the beginning; later it all unravels.

The audience is no better at this time game. With the first speaker they are fresh, and they ask questions as if they had all day. Most of the questions are designed to show how intelligent and informed they themselves are. You sometimes wonder why they deign to hear the talks. Perhaps they should be giving the talks themselves, and from the sound of the questions, maybe they are.

Even as we unconsciously prepare talks that are too long, we rationalize that we will be able to adapt the talk on the spot to whatever time conditions seem to prevail. This is, however, a delusion. Think back to the occasions that you have seen a speaker told that his talk must be made shorter, or must be finished up in some small amount of time. What is the speaker's reaction? Does he take his pile of transparencies and put most of them off to the side? No way. Whatever time must be saved, he apparently imagines that he can gain by speaking faster! It seems impossible to derail the prepared litany of subjects. Every chart must be shown, even if the speaker has to sound like Donald Duck in a frenzy of time compression.

But again I digress in philosophy while the afternoon has waned. It is finally my turn to speak, but the entire quota of time for the agenda, as well as the audience's patience, has long ago expired. Several principals have already left, mumbling the perennial airplane-to-catch excuse. The others are glancing at their watches and yawning.

Nonetheless, I lug my 40-minute pile of transparencies to the head of the table. I wasn't ready for prime time anyway.

Credentials: Who Needs Them?

My biography is being read from the podium. I am thinking about the speech I have to give, not really listening to the introduction. It is just as well; my biography bores even me. When patches of it thread through my listening defenses, I cringe inside. "That's not me," I cry out mentally. Of course, I wrote it myself, following the usual conventions of our profession. That is why it is so dull.

When I read a good paper — or a particularly bad one — I turn to the back of the journal to find out something about the author. But my curiosity about the person nearly always remains unsatisfied. All the biographies are the same: the authors went to this school or that, took one job or another, and currently hold such and such esteemed position. They also served on committees that I have never heard of. In some journals there will be a picture, a mug shot with bulging eyes. I imagine the caption, "... currently serving 10 to 15 years for armed robbery." Is this the real person? Is that all there is? Or are we so hung up on our professional facades that we have forgotten to be even a little bit human?

When you read a book of popular fiction and the flyleaf has a paragraph about the author, does it ever say that he or she served on this committee or that? Never. Usually it says something to the effect that the author lives alone on a windswept, deserted beach with his large Doberman named Cerberus. When not writing he spends his time playing the harpsichord and tending to his collection of Ming vases. Perhaps the back cover has a picture of him walking the beach, tie askew, in an old tattered sweater. His weatherbeaten face appears thoughtful. Now there is a real person, you think. Even his dog looks cerebral, his head, near the lens, interestingly distorted.

The biographies in the personal advertisement columns in the backs of some popular magazines are another extreme. Here typical advertisers, seeking companionship, are incredibly rich and do not seem to know quite what to do with all that money. Naturally, they are handsome, trim, well-educated (usually a physician, seldom an engineer), world-traveled, and talented in something like music. The advertisers go on, trying modestly to describe their many virtues. They are usually looking for someone to roam the world with them, wind-surfing and romancing their way together toward an endless sunset. They make me sick. I ask the same question — are these people real?

But is there no way to enliven our own dull professional biographies? I suggest several possibilities. The problem is that we hide our own natures and abilities too well. Have you ever found that someone you worked with for years professionally had some outside talent that you were completely unaware of? Perhaps this one trained originally as a concert pianist, or that one worked as a professional magician, or ran illegally in the Boston Marathon, or was an avid participant in hang-gliding. When such a revelation happens to me, I suddenly see the person in a new light. The things done professionally take on a new character and flavor. The whole person is somehow shaped.

But not all of us have secret talents. Thus we might allow a special waiver in biographies for the untalented, with everyone entitled to just one lie. That rule would give free rein to expression and creativity. I can imagine one author saying that he conceived the idea behind this paper on compound semiconductors while working as a mercenary in Central America, or another confessing that she felt compelled to write down her feelings about computer vision while living as a bag lady on the New York City streets. Another might claim to have given up a career in major league baseball in favor of research on architectures for distributed processing. Yet another might cover any deficiencies in the presentation by adding the caveat that the paper was written while the author was recovering from an accident that had caused a considerable amount of temporary brain damage.

Another alternative to enliven biographies would be brutal, perhaps even uncalled-for, honesty. "The author has no talents whatsoever, and is fortunate to be able to work at this boring job." "The author left his most recent job at Megalithic Megatronics last September, when he was fired for incompetence." "The author wanted to be a movie star, but was unable to get a screen test, and in desperation turned to this silly work on computer networking." "Because she has never been asked, the author serves on no IEEE committees." "The author has no advanced degrees, since as a condition of receiving his BSEE in 1972 from Upper Transylvania University, he was warned never to come back."

If we all had impressive credentials, then it would probably be unnecessary to enliven our biographies. As a case in point, I remember the day years ago when as an editor I received two papers for inclusion in a special issue. I had the two authors' biographies side by side on my desk. One was a page and a half long, the other only two lines. The one told of the many committees on which the author had served, the journals for which he had reviewed papers, and the conference sessions that he had chaired or organized. It was most impressive. The short one merely stated that the author was a Hero of the Soviet Union, and had received a Nobel Prize in physics. Poor fellow — not much to write about.

Perhaps the only thing better than having superb credentials is having no credentials whatsoever. I was in the audience at an IEEE conference for a talk by a friend, when it was announced from the podium that the author had been unable to attend, and had instead sent a substitute to give his paper. Naturally the substitute did not deserve, or receive, an introduction. I was calculating my chances of escaping unnoticed when I saw that the person mounting the podium was actually my "absent" friend, the author of the paper. He delivered the paper authoritatively in the third person, occasionally commenting upon the work in an editorial way, and not always favorably. I heard the people around me, who of course did not realize what was happening, whispering excitedly about what a marvelous talk it was, and how effective it could be to have a talk presented by a disinterested, but knowledgeable, associate.

I was fascinated by the event. To be free of one's persona, of one's ego, of one's stake in the audience's acceptance of the paper, that was such stuff as dreams are made on. To have the weight of responsibility for self lifted away, to be able to praise or criticize one's own work — what more could one ask? Perhaps I shall try that approach one day. Meanwhile, if we are forced to stand by our biographies, let us at least try to make them more interesting. I myself will try, but that may prove more difficult than writing my next paper.

A Way with Words

How beautiful language is! The possibilities for expression are limitless. The dictionary contains a sea of half a million words that can be strung together in sequences to achieve infinite degrees of nuance in meaning. However, in our daily spoken language we generally confine ourselves to an impoverished pond of less than a thousand words. In my laboratory and its environs, the ten most commonly used, in order of frequency of occurrence, are:

the
of
and
competitiveness
to
infrastructure
a
in
paradigm
that

Naturally this list changes slowly with time. Until a few years ago I had never once in my life used any of the words "competitiveness," "infrastructure," or "paradigm." Those words resided undiscovered in my dictionary, waiting for the unique societal circumstances to arise that would require their use.

Now their long, mute vigil has been rewarded with the linguistic equivalent of 15 minutes of fame; today it is impossible to utter more than five sentences on the subject of technology or technology policy without using one or more of these indispensible words. I admire the foresight of our ancestors, who knew that some day we would have such compelling need for them.

It is a mystery how these words get started. Who first plucks them from the muddy depths of language and casts them into the mainstream? This is obviously a tricky business. Suppose I ventured into the dictionary and discovered some word like "nonnipotent," meaning "having power over nothing," which should be generally useful in a variety of policy situations and IEEE meetings. I could start using it in writings (like this) and speeches. Maybe it would take. "He's got something there," people would think. On the other hand, they might well be muttering, "He's an idiot." You take your chances in this uncharted territory.

But to be a contributing member of our technological society today, you have to know how to use heavy-hitting words like "competitiveness," "infrastructure," and "paradigm." At first you may be tentative. That is bad, and will count heavily against you.

To gain experience in using them, it is best to practice in the privacy of your home. For example, you might mention to your dinner guests that you have used a new paradigm for the cheesecake. If they frown and glance uncertainly at the cake, then probably you have not used the word with maximum effectiveness. Perhaps you might mention reassuringly that you are working on a new infrastructure in your basement, the old one having crumbled.

When uttered properly and in the correct context, these special words will be seen to have dramatic impact. The listener should nod knowingly with an expression of rapt attention, imparting a sense of deep concern and shared intelligence. In fact, it is best to work on your own response to these trigger words, since the speaker will undoubtedly be watching you closely to gauge your worthiness for such elite conversation. Be on the alert, however, for impostors, who might either be using these precious words carelessly, or as a test. If you are seen to nod knowingly at the wrong time, people will substantially lower their estimates of your IQ.

The one word that it seems impossible to misuse is "competitiveness." This is because no one actually knows what it means. However, it reminds me of the old Groucho Marx television show, where if the guest happened to utter a "magic" word, a duck would drop down from the ceiling carrying money in its beak. Today's magic word is "competitiveness." If you are seeking money from some government agency, then be sure to use it. Whatever you are doing, say that it is "critical to competitiveness." Be sure to use the adjective "critical," since this is the current hot modifier.

As an example, suppose that you are attempting to get funding for a mathematical analysis of the wave equations where the boundary conditions are in the shape of a "Z." You must say that this study is "critical to the nation's competitiveness." Since no one has any conception of what factors actually affect competitiveness, no one will dispute your claim.

You might amplify by mentioning that other nations have already completed analyses of "X" and "Y" shapes, and that the foremost competitor is rumored to be working on the critical "Z." At this point glance heavenward and allude to the possibility that a whole new infrastructure for alphabetic resonators is in the process of formation. The analysis of the "Z" will likely create a dramatic shift in paradigm.

If you say these words with sufficient confidence, you will be rewarded by seeing the faces in your audience turn suddenly grim. There will be a surge of feeling in the room of shared national incompetence. The decision-makers present will do their part by granting you whatever funds are necessary to close the gap on the "Z" analysis. This will make them feel good about their own roles, and they will rush to their next meeting to hear about something else that is critical to competitiveness.

The awful truth about this sort of behavior is that you end up believing in your own technospeak. Once having used these potent words, you become awed by the criticality of your work to the global economy. It is hard to sleep with the "Z" analysis unfinished. Carrying the weight of the technological world on your shoulders is a heavy burden. As you catch sidelong glimpses of yourself in passing mirrors, you are impressed with your new bearing of importance. You owe it all to the mastery of these latest words in technospeak.

A warning, though. By the time you read this page, the magic words may have changed. Do not be caught using obsolete words. You will be put on a list of people never allowed to receive funding again.

The Papermill

A long time ago I wrote my first technical paper. I was really proud to see my name on the index of an *IEEE Transactions* issue. I was an author! Many people were impressed — me, my mother, and possibly a few friends and neighbors who happened to glimpse the issue as it lay casually placed on top of the living-room coffee table. For years to come I would be able to drop into technical libraries and find myself incarcerated, as it were, in the dusty stacks. What a marvelous system for publication we engineers had!

In the decades since, I have had occasional glimpses of the Truth. I had one vision of it in the religious quiet of the libraries of the British Museum and of Cambridge University in England, where the actual notebooks of such immortals as Isaac Newton made me shiver with the shared pride of a great scientific heritage. I've marveled at the lasting and renascent wisdom that can still be mined from our great papers — Nyquist in 1928, Rice in 1948, Shannon in 1949, to name a few in the communications field. Indeed, we engineers have much to be proud of in the record we have left behind.

However, there is another vision of our publications system. It is the myopic, day-to-day vision seen from the participants' perspective — reader, author, reviewer, and editor. From their viewpoints it is indeed a miracle that there is any trace of a legacy to pass on to the next month's engineers.

Let's begin by dispensing with the myth that the publications are for the readers (*Spectrum*, of course, is an exception). Naturally, readers are a necessity, but we must be careful not to overestimate their value. For one thing, readers are a nuisance. They continually complain about the worthlessness of the material. They want more practical papers, they say — not these incomprehensible theoretical exercises. Should the journal be so fortunate as to publish one of the practical papers it so rarely receives, the readers complain about the noticeable drop in quality.

Worst of all, the readers make the outrageous claim that they are, in fact, readers. They line the shelves of their offices with reams of pretentious journals, proclaiming to the world their erudition. But I've heard the estimate that the average technical paper has approximately three readers.

It's hard to evaluate this estimate, because there's some kind of uncertainty principle involved in studying readership. The process of measurement affects the system. People will fill out polls saying that they read *n* papers in an average issue: *n* is never equal to zero. No one would admit to not reading a journal that is received. Thus, there is some kind of unspoken agreement that among us engineers *n* equals, say, three. This means that, for example, in the *IEEE Communications Transactions*, with about 15,000 subscribers and 15 papers, the average paper would be read by 3,000 people. If you believe that, you probably also believe that giant vacuum tubes once walked the earth. On reflection, however, any author should be quite pleased to have two or three good readers who would study the paper and pass on its essence to their associates, for this is the way our system truly works.

The author, on the other hand, feels that his or her paper is a great masterpiece written in deathless prose for consumption by the masses. His or her ego is on the line when the precious opus is dropped in the mail for the cruel scrutiny of the review system. The typical author hopes for a telegram from the editor the next day, telling him his paper is so important that it will be published immediately, without review, in a special issue of the journal with his picture on the cover. As the days tick by without reply the author's apprehension grows. Has the paper been sent to his arch rival for review? Will its simplicity be apparent, or has it been well concealed by the usual practice of beginning with the general case, and only explaining special examples (where the paper should actually have started) as afterthoughts treated as if they were of no importance to the potential reader? The euphoria attached to the initial generation of the paper dims, to be replaced with a latent anger ready to be fired at hostile reviewers. Meanwhile the author's energies are consumed in preparing the next paper, entitled "... Part *n*," "Return of ...," or "Further Implications of ...."

While this is going on, the reviewer has received the paper, with a request from the editor for an opinion. As befits the dignity attached to rendering such a judgment, the reviewer first hurriedly turns to the back of the paper to see if he himself has been referenced. Probably he has been, since this is how the editor generally selects reviewers, but heaven help the author if he has neglected to refer to an important work published by the reviewer.

I remember being attacked during a patent trial on the basis of a paper I mentioned in a textbook I coauthored. Why, the barrister asked, did I refer to that paper if it in fact had no bearing on my invention? I tried to explain, "When one writes a book like this, one tries not to make any more enemies than one can. General procedure is to refer to everybody in sight; it helps sell books." There were no more questions along this line.

Next the reviewer turns to the front to see who the author is. There are three possibilities. First, it can be good-friend Allen, who is known never to produce a bad paper, by definition, since it would be professional death to pronounce one of his papers bad. Second, it could be well-known-faker George, who is trying to slip one through the review process again, hoping to get a less astute reviewer than yours truly. And third, it could be that rarity — the unknown author. (Incidentally, there is always an outcry about removing the author's name from the manuscript: this procedure only delays by milliseconds the recognition of the above categories.) In the case of the new, unknown author, it will be necessary for the reviewer to put in a certain amount of work in order to establish his or her own preeminence in the field. After all, this author should realize just who it is he or she is dealing with.

Finally, in the middle of all this, is the editor. The main skill necessary here is that of translation. The letters from the reviewers and authors must be translated before transmission. "This is the worst paper I have ever reviewed ..." becomes "Reviewer A has found a certain deficiency in your paper ..." In trying to placate these warring factions, the editor makes few friends. Most of his letters are apologetic. "I realize that the two years that have elapsed in the review of your fine paper may seem exorbitant to you ...," he begins. Untouched on his desk is the yearly paper proving Einstein's theory of relativity wrong, and the slew of letters from would-be authors protesting that, being outside the "system," their revolutionary papers will not be accepted by the traditional reviewers. In the back of the editor's mind is the nagging doubt that maybe they're right. The trouble is — it just isn't likely.

Indeed, it's a merry-go-round, and few of the papers published this year will be mentioned in the year 2000. We could debate the relative worthlessness of individual papers. Certainly there are layers and layers of worthlessness, but it's hard to say from moment to moment which papers will in the years to come merit special reconsideration. The system may not be as fair as it should be, and everyone concerned will from time to time have legitimate complaints, but I for one think that time has proven its value to all of us. On the whole, it works.

Part 3

The Fickleness of Technology

A little more than ten years ago I wrote my first column for *Spectrum*. Entitled "Rampant Technology," it was about how fast everything was changing in the electronics field. Unlike most of the other essays collected here, which are largely about the timeless social drama of our field, this first column seems old and rusty. The devices that I mention — 64kb DRAMs and the Motorola 68000 — have long since been superceded, and their mention lends a sepia-colored overtone of antiquity to the subject matter. Yet that nuance in itself proves the point; things have changed so much in that decade that this column now reminds me of some musty, deserted laboratory of yesteryear.

The concept of continuous change is so rooted in my engineering mentality now that I have returned to this theme again and again in my columns through the years. Yet there was a time when I was blissfully unaware of this intrinsic truth of our field. Looking back to this time of youth and technological innocence, I often think that electrical engineering pulled a kind of bait and switch tactic on me, luring me with dreams of enduring competence and only then letting me know that the garbage heap of technological obsolescence would be breathing down my neck the rest of my career.

As a child I used to listen to radio shows on my old tube-type radio. In my darkened bedroom I could see the filaments of the tubes glowing orange, and I would imagine that the voices of the Lone Ranger and the Shadow were generated in those warm, dancing glows. Someday, I thought, I would know how those tubes worked. Someday I too would be a keeper of the electrical secrets.

When I was a teenager, and even all through college, I thought it was possible to learn essentially all there was to know about EE. Sure, there might have been some little area that you didn't know well, but you had your *degree* and you were certified as knowing this stuff. What I had in mind was rather like the master-apprentice system practiced by all the unions of skilled craftsmen. You served your apprenticeship and eventually you yourself became the master — the keeper of all wisdom and the practitioner of consummate skill.

I remember buying my engineering textbooks when I was in college and thinking how they would serve me the rest of my life. When I left with my degree I packaged my books carefully, since everything that I would need subsequently would be in there somewhere. I still have them. I don't know exactly where, and it does not matter; they are now useless. My dreams of perfecting my skills at designing vacuum tube circuits have long disappeared in the smoke of transistors, integrated circuits, microprocessors, lasers, and so on, and so on. Nobody ever warned me or gave me any clue that what I was learning would be so quickly obsolete.

In the movie *National Lampoon's Vacation,* Chevy Chase's character drives his family across the country from Chicago to "Wallyworld" on the west coast. Packed into the station wagon with Chase and his wife are their two kids, an aged aunt, and an ill-tempered dog that nobody seems to like. At a rest stop alongside the highway the dog is leashed to the back bumper. A little time after leaving the rest stop, speeding down the highway, Chase is pulled over by a state trooper with tears in his eyes. "The little dog ran as fast as he could," says the trooper through his tears. There is nothing left but an empty collar tied to the bumper. Sometimes that is how I feel about technology. I run as fast as I can, but the whole field moves ahead of me so fast that, like the poor dog, there is nothing left of me but an empty collar.

A number of my essays harbor the yearning for a more stable world. Even in life in general I often ask myself why it is necessary for things to keep changing. What if I like it the way it is now? Too bad, I answer myself. Things may get better, or they may get worse, but the only guarantee is that they will be different. In the world of technology, in contrast, there are certain guarantees. True, things will get different and what you know now will be useless tomorrow, but the differentness will have certain predictable attributes. For the engineer the world of tomorrow is guaranteed to be more complex, to be faster, and to be smaller.

Shrinking Components

It is also about a decade since I wrote "The Incredible Shrinking Transistor." Even that essay seems to have been written in an era of technological innocence. It is enough to make me feel that I should never mention anything about technology per se in anything that I write. I am sure that any IEEE Fellows who wrote articles about the future in the commemorative issue of the *Proceedings of the IEEE* (May 1962) mentioned in my column will share my feeling. This issue is a prototypical example of a well-known engineering trait when looking at the future — we are too optimistic about short-term progress, and we are far too pessimistic about long-term progress. It really is incredible that no Fellow in writing about the future in 1962 saw the impact of the integrated circuit. Yet it had already been invented at that time!

The Fellows of the IEEE were not the only people having a hard time coming to grips with the idea of a constantly shrinking transistor. Some of us old-timers in Bell Labs remember the famous RSI devices that our own semiconductor people were pushing on us in the early 1970s. RSI, lest there be any doubt, stands for right scale integration — not *large* scale integration or *small* scale integration, but *right* scale. The argument was that if transistors got too small, they would be too hard to make and thus be uneconomic. Somewhere there was an optimum size. I remember even at that time that many of us were skeptical, believing that right scale meant that that was all we were capable of doing in our own facilities. Anyway, probably the argument failed on the principle that people forgot that the transistors were less important than the fact that all the wiring was also included in the circuit. Ultimately, the wires would be what mattered.

Soon after the early integrated circuits were developed, Gordon Moore, one of the founders of Intel, conceived of his famous curve. Known as the Moore chart, it said that transistors would get smaller at an exponential rate. The curve of transistor size versus time was a straight line on a logarithmic plot. Every engineer knows this law; it lies at the very heart of our profession. Constant, exponential progress — wonderful yet ominous.

Moore's curve fits more than just transistor size. It seems to characterize almost every parameter of engineering achievement — circuit speed and density, transmission rates on optical fibers, etc. Furthermore, the exponential progress is not merely a transient phenomenon. It seems to continue over decades in spite of all the dire predictions by experts that the curve will flatten. In the case of transistor size, there have been many projections through the years that the size of transistors would have to stop shrinking for this or that reason. For example, people said that we could not make photolithographic masks below one micron in size because this would be smaller than the wavelength of visible light.

Experts now say that transistor sizes will stabilize at around 0.1 micron line width. Maybe they are right. The current arguments are not particularly esthetic, but they fit into my "bugs under the rock" theory. This is the way I always think about things that just get too messy to handle. I guess I had a lot of experience in my youth turning over simple-looking rocks and seeing all the awful things that crawled out. In the case of 0.1 micron line widths we have a multitude of dreadful, messy things happening — electromigration, tunneling, heat dissipation limits, etc. Awful things crawling out.

Whether or not the current crop of experts is right about the eventual stoppage of transistor shrinkage around the turn of the century, it seems to be a law of nature that technology requires an exponential increase in functionality. Someone told me that soon after he conceived his famous curve, Gordon Moore was concerned that engineers would not be able to use all the functionality of the increasing density of circuitry. As it turned out, it was no problem. The shrinking transistor is the furnace in the basement heating the landscape and powering technology. When it stops shrinking, we will have to come up with some other way of fueling the technological economy. Functionality must increase exponentially. Always. Someway.

The Blurring of Reality

With all this shrinking going on I have this sense of a loss of physical dimensions. Everything I know is going down the drain, while I stand by helplessly and watch. Like a lot of people, I guess, I am more comfortable with things that I can see, touch, and visualize. Give me a wrench and let me crawl under the car, and I can look up through my sweat and see that the thigh bone of the car is connected to the ankle bone, or whatever. But show me an integrated circuit these days and I can only shake my head sadly. Who knows what's inside there?

Years ago I saw some television drama about a World War II spy who parachuted behind German lines to get a look at a new radar installation. He took with him a scientist (never an engineer, I'm afraid) who was a radar expert. After much derring-do, the pair got into the radar installation. While the professional spy looked on helplessly and worried about how much time they had, the scientist opened the radar cabinet and a sheet of paper fell out. "What's that?" asked the spy. "Oh that's nothing," said the scientist, "just the circuit diagram." Frantically the scientist traced the physical wires with his hands. The circuit diagram rested unseen on the floor; they left without ever looking at it. The scientist tried to mumble technical-sounding stuff about the wires. I turned the TV off in disgust.

I suppose there were days when you could look at wires and components and guess what something did, but those days are long gone. Can you imagine a scientist (or engineer!) breaking into a computer complex today and trying to figure out what it does? Lots of luck! You wouldn't have a clue.

After the publication of "What's Real Anymore?" I got an E-mail message from Chap Cutler at Stanford. Cutler said that he had a class of undergraduates that was learning about electronics. In order to get a physical feeling and intuition about the subject, they took a television set apart, component by component. They cut open a filter capacitor, and broke open the picture tube to inspect the electron gun, and so forth.

Cutler's class finally got to a large integrated circuit. How were they to see what was inside? They placed the circuit in a small vise and focused a microscope on the chip. The microscope was attached to a television camera and display so everyone could view the dissection of the IC. They then took an ultrafine file and began to file away, ever so carefully, bits of the plastic package. At first only plastic surface was seen on the television display. Later still, only plastic was seen. After much careful filing, they began to see the outline of the leads at the bottom of the package. As the last of the plastic package and the leads disappeared under the file, the realization began to dawn on the class. *There was nothing inside the chip!*

Cutler told me in his mail message that the class had begun to worry that the integrated circuit revolution was a hoax. Perhaps all the elaborate preparation of mask designs, etc., was only a cover-up in order to sell empty plastic packages. Cutler himself seemed very concerned. Of course, he has also been known to kid around at times.

In recent years the enrollment in electrical engineering has decreased, particularly relative to mechanical engineering. My personal theory is that the disappearance of our field into virtual nothingness is partly to blame. Electrical engineering isn't as much fun as it used to be.

Meanwhile mechanical engineering continues to grow in popularity. There students can see and touch things. But it is difficult for electrical engineers to sit back and watch this displacement of our natural domain. Thus it was a welcome inspiration when a group of electrical engineers came up with the idea of microminiaturizing mechanical components using photolithography, the same way we make microelectronic circuits. This would shrink the mechanical components out of existence also. Let the mechanical engineers worry about things they can't see, feel, or smell. What a wonderful idea!

The idea of the photolithographic production of microminiature mechanical components was indeed a revolutionary concept, and it was featured in many newspapers and magazines. I wrote the column "An Engineer in the Land of Lilliput" about my experiences with a very tiny wheel, about the diameter of a human hair. It is an amusing story, which I hope you will enjoy, but after its publication I discovered once again that not everyone appreciates such stories in the vein of humor. I got several letters from people admonishing me for what I did, or even for talking about what I did, if I didn't actually do what I said I did — or whatever. I will let the ambiguity persist, and apologize to anyone who thinks I misbehaved. My only excuse is that I try not to take myself too seriously, and I try to let my audience take what I say and write in that context.

The Demise of Heathkit

Someone sent me a newspaper clipping the other day about Heathkit going out of the electronic kit business. A few weeks later it was a front page story in the *New York Times*. The story told about people like Barry Goldwater who used to fly up to Benton Harbor just to stock up on the latest Heathkits. Other people were mentioned who collected a hundred or more kits so they would never be without a project.

I used to love making Heathkits myself, and the essay "Goodbye Heathkit" is my lament for what is no more. There was something therapeutic about following the exacting instructions of cutting and soldering wires. And there was nothing quite like the thrill of the so-called eureka of turning on the television or hi-fi and seeing it work. Moreover, people were always impressed that you built whatever it was yourself, and you had a comprehensive manual in case anything ever went wrong. Finally you told yourself that you had saved all kinds of money. The problem was, that was never quite true.

"The Electronic Hobbyist" is another lament for the passing of electronics as a hobby. I suppose it is too much to hope that your professional work could also be related to a hobby. As they always say, that is why it is called *work*. I have gotten a number of letters from old-timers about how they miss the old electronics, the smell of the soldering iron, and the sense of accomplishment that comes with running wires and turning screws. Even some of the younger EEs have built electronic equipment in their youth, but they are the last generation to do so.

I used to enjoy tremendously electronic flea markets. In its early years the Trenton Computer Fair was a marvelous way to spend a fresh spring day in April. I loved to pick through old junk boards and piles of used disk drives. I am not even sure why this was so much fun; I certainly didn't need any of it. But occasionally I would buy something for almost nothing and actually make it work. What a thrill that was! I knew EEs that just collected old electronic parts for no discernible reason. I wasn't that perverted, but I know the feeling.

Alas, all of that is gone. It is much, much less expensive to buy a whole circuit board than it is to buy the parts and wire the board yourself. The vast majority of the wiring is sunk into microscopic chips where no one can get at it, and no one cares anyway. Today they still have some ancestor of that Trenton Computer Fair, but all you have is hundreds of commercial booths selling the same prepackaged programs and clone computers. When you see the prices of motherboards and other computer boards coming to the dealers from the Orient, you realize that electronics in the small has been reduced to a worthless commodity. Sigh!

Today when I buy a piece of home electronic equipment, like a television set or a hi-fi, I buy it with the realization that when it breaks, no one will fix it. It will be thrown away, and I will buy another one. I used to fix my own television sets. People were really impressed, especially me. The last two times I had a television break, I took it to a repair shop. One shop charged me $75 to replace a "high voltage capacitor." The set originally cost only $85, and I got the circuit diagram and found that there was no high voltage capacitor. They had replaced a ten cent diode in the power supply.

The next time my television failed, a different repair shop replaced the "high voltage transformer" — an expensive item at $80. The clerk shook his head sadly at my bad luck at needing such a high ticket component. Later I had the occasion to repair the set myself, and found that the transformer was the original one, welded onto the frame. A few months ago I moved my residence, and I put two large television sets out in the street for the garbage. No one even picked them up.

Living on the Fringe

Since technology has taken us to mystical places, I feel less sure of my role in society as an engineer. Engineers make things, I keep telling myself. But do we really? As reality fades behind the computer screen, the concept of "making things" becomes more obscure. Most of us do not really make things at all; we deal with information that is eventually used in some abstract way to make something or other. Sometimes I think of us as living on the fringe. Fortunately, society is relatively rich and can afford a lot of people working at the fringe, not quite producing anything of substance, but somewhere a long way down the food chain.

Apparently I was worried about the role of engineers in society when I wrote "The Paper Airplane." I can still see that paper airplane floating out of the murkiness of a technical meeting with its terse criticism of what we were talking about. I imagine a concerned layman sitting in on one of our conferences and thinking to himself about how it is that people get paid for discussing these esoteric subjects. I mean, I think that way myself sometimes. From time to time I feel guilty about it, but then I go back to feeling underpaid. Someone has to do this abstract work, after all.

It wasn't always this way. In fact, it only got this abstract in very recent years. The role of engineers used to be much simpler to define and measure. One of the great novels that describes an engineer of yesteryear is Mark Twain's *A Connecticut Yankee in King Arthur's Court*. Mark Twain's engineer knew all there was to know about how to make things. After being mysteriously transported to the Middle Ages, he was able to singlehandedly recreate much of technology, albeit under the guise of magic. This has always been a great fantasy for me — the idea of going back to a time when your special knowledge has a unique importance, as compared with today's world when there are a lot of us and nobody cares particularly what you know about programming digital signal processors, for example. The world couldn't care less.

In "A (Very Modern) Connecticut Yankee in King Arthur's Court" I play out this fantasy with an engineer from today. I had fun with the essay, but the message here is a disappointing one. I think any of us would have an extremely hard time impressing King Arthur with our magic. The stuff we work with these days is too complex and too abstract. For any of us our knowledge ends somewhere in the hierarchy of a system. Many electrical engineers do not understand the depths of physics in today's devices, whereas the more physics-minded people haven't a clue about the higher layers of the system in which their device is embedded.

Only a very modern society can afford people living on the fringe. In a number of works of fiction some calamity returns the world to a more primitive place where people on the fringe cannot be supported. Usually this is a nuclear disaster. After an atomic attack, there is not a great deal of need for lawyers and investment bankers. In *Lucifer's Hammer* by Larry Niven and Jerry Pournelle (the *Byte* columnist) the calamity is a comet that plunges into the ocean, causing tidal waves that wipe out organized society. Groups of people band together in enclaves, trying to recreate agrarian societies while fighting off marauding bandits. An ungainly computer programmer, Dr. Dan Forrester, approaches a California enclave where he is halted by the sentries. In order to be allowed to join the enclave, Forrester must offer some needed skill to the group. The question is what can he do that is important? Writing computer programs does not count.

I think about Forrester's dilemma as a modern equivalent of Mark Twain's hero. What could you offer to the enclave? Have you any knowledge or skill that is marketable in a basic society? Or have you, like me, lived too long on the fringes of a rich and complex society? In

Forrester's case, as it turned out, he had much to offer. As a token of his worth, Forrester showed the sentries a precious book wrapped inside four protective baggies — *The Way Things Work, Volume II.* The sentries wanted to know what else he had. "The 1911 *Britannica*," Forrester said. "And an 1894 book of formulae for such things as soap, with a whole section on how to brew beer starting with barley grains." Needless to say, he got in, but his most important knowledge was yet to be shown — he knew how to make mustard gas.

Setting out to work for the enclave, Forrester had a grand goal. "We have to save the power plant," he pleaded. "We can rebuild a civilization if we have electricity." Any electrical engineer can relate to that, but how many of us could rebuild a power plant? As an EE in a ruined society, that would be your first job. Get the lights on. Could you?

The Rise of Complexity

Now back to the present. Someone else can get the lights on. Someone else can fix the television. That is not exactly what electrical engineering is all about anymore. We deal with a world that is steadily increasing in complexity. It is all we can do to escalate the power of our tools, our concepts, and our models in order to keep pace with that complexity — much of which is our own doing, of course.

"Coping with Complexity" is the most serious of my essays collected in this volume. For that reason I like it less than most, but occasionally someone should worry philosophically about what is going on. I am particularly concerned for the armies of programmers working on million-line programs. To me it sounds like being the third rower from the left in an ancient slave ship. You work in excruciating conditions pulling away at your oars, but no one even knows that you exist. Collapsing in your chains, you are pulled away and replaced with another nameless individual. It is probably not that bad, I hope.

I wish that there were a powerful theory of complexity, but then I guess that would make it not complex anymore. There are a lot of important systems whose behavior is largely unpredictable because they are too complex to understand — telecommunications networks, the weather, economic systems, traffic, and so forth. Perhaps they can all be explained by chaos — or is it fractals this year? I do remember that long ago I attended a conference that was going to explain all complex systems in terms of hierarchical structure.

If I had paid attention at that conference, perhaps I would have learned some great lesson, but all I can remember is the numbing lesson that I got in humility when I was introduced to the great Princeton economist, Oscar Morganstern. When Professor Morganstern was told that I was speaking about information theory he glanced briefly in my direction. "Shannon ... couldn't make it?" he said with raised eyebrows in his deep, academic voice, whereupon he turned and walked away. Everyone in the room looked at me in the deepening silence that followed his exit. "What is *he* doing here?" they seemed to be saying. I wasn't sure what I was doing there either. Not your greatest confidence builder.

Along with the rise in complexity and the trend away from the physical world towards the virtual world, there has been an increased demand for software. Get rid of the physicists, they say. All we want are more computer scientists to write code. Better yet, the ones we have should write more code, so we don't need any more people. That cry for more efficiency in the generation of software is the theme of the essay "Software Productivity." In contrast to the column on complexity, this one has less philosophy and more amusing images. I like it better.

Fred Brooks' likening of the silver bullet of vampire fame to the problem of software productivity is one of the best known metaphors in our profession. He put this idea into his Turing Lecture for the Association of Computing Machinery, but unlike the ideas heard in most award acceptance speeches, this one caught fire. Is there a silver bullet or isn't there? The question resounds everywhere. Periodically I hear about some heavily promoted program to find the silver bullet. Software productivity will be increased by a factor of three in the next five years, says the announcement. The time goes by without a great deal of further fanfare. Then another announcement goes out. Software productivity has been increased by a factor of three, says the new announcement. People shrug their shoulders. Who would know whether it has been increased or not? (Although I feel bound to add that the Software Engineering Institute in Pittsburgh keeps statistics and guidelines about such matters.)

Meanwhile those million-line programs lumber around like clumsy dinosaurs, either a deadend accident of evolution or a new breed of living creature. After you have put a hundred million dollars or so into a piece of code, no one wants to abandon it and start over. Yet the old code is such a mess of spaghetti that no one can work with it any more. So you keep adding features to the hopeless old code. As a consequence, the program becomes even more unworkable, even as the investment becomes greater.

Maybe someday in the distant future archaeologists will discover the fossilized remains of our giant programs. They will be reconstructed in museums, and kids will come in to stare at the millions of lines of archaic code. The plaque will say that the program was written in an ancient language called *C*. Alongside will be a picture of some contemporary scientist who was the first to decipher this dead language. The picture shows the scientist standing next to a large stone containing inscriptions of both a C and a FORTRAN program. The little kid looks up at his mother. "Did big programs like this really exist?" he asks. His mother doesn't know much about archaeology. "I think it was the ice age or something," she says. "They just disappeared from the face of the earth." The little kid shivers. Imagine these giant programs walking the earth! What a scary place it must have been.

Hardware Isn't Easy Either

Not everyone designs million-line software programs. After all, software has to run on something. That is what hardware is for. Strangely, the complexity of hardware continues to grow even as the boxes themselves shrink. Today a lot of hardware design is plugging together small chips of infinite complexity. Whole books get written about one little chip, about what signals it likes and what it doesn't like. Furthermore these chips are very fussy. You have to keep them happy in many complicated ways. Thus the designer spends most of his or her time trying to find out about the care and feeding of the chip. None of this "give me a soldering iron and I'm off" kind of thing.

"What's Bugging Us" is a lament about the frustration and lack of respect engineers get in trying to debug circuits. It contains some daring new ideas about bugs in electronics, building on the pioneering work of Louis Pasteur. Unfortunately, even if the right vaccines could be developed, they may not be timely because the nature of hardware design seems to be changing out from underneath us. More and more designs are rather standard, based on microprocessors, special-purpose processors, and support chips. You get these circuits from the manuals, and all that is left is the glue and the software. And nobody cares about the glue.

There are, of course, engineers who design the VLSI chips themselves. We crossed an important threshold some years ago when it became faster and easier to design (or simulate) custom VLSI chips than to build a breadboard out of small scale circuits. Unfortunately, this leaves my Heathkit mentality out in the cold. Can you imagine a VLSI design kit? Yet most of the advertisements today in *Spectrum* and other electronics publications are for software programs that do circuit and system simulations of various sorts. Now you know how the circuit will perform before it is fabricated. This takes all of the fun out of it. Somewhere there must be real people, working with real physical circuits, rather than crouching behind computer screens, but they are getting harder to find.

In case there is any design freedom left, the world has decided that everything must conform to standards. I am very ambivalent about this. On the one hand, standards stifle innovation and give the competitive advantage to nations and companies that can mass produce commodity products, rather than to those that can innovate. On the other hand, as a consumer and ultimate user myself, I demand standards. My greatest frustration as a military advisor was with systems that had used proprietary solutions. Years later you find yourself with a computer that no one supports, with software written in a language for which there is no compiler, and so forth. Yecch!

My ambivalence about standards comes out in "Plugging In." Why does nothing plug together like it should? I don't even understand today why the three different keyboards that I use everyday —workstation, home PC, and laptop — all have their control keys located in different positions. Surely they could get together on this. As it is, I keep hitting "caps" by mistake even as I type this paragraph. If we cannot even get something like a keyboard standardized, how are we ever to get electrical and software interfaces to work together?

In "Plugging In" I wrote about my long-standing reticence to plug electrical connections together. I might have heeded my own advice, as last week I had yet another close encounter with dangerous connections. I had just purchased a new high fidelity amplifier and was inordinately proud of the "deal" I thought I had gotten. But pride, as we have learned, goes before a fall. Grand debacles are usually precipitated by small, ordinary events. In this case it was my son asking if he could use my PC to play a computer game. Of course, I said, if he would wait just a minute for me to plug the computer's audio output into my new amplifier.

Now another old maxim comes into play — haste makes waste. Any self-respecting EE must make simple audio connections quickly and with a certain professional flair. Unknown to me, however, the computer's audio output was very "hot," whereas the digital volume control on the feature-rich amplifier had no external indication that it was turned on high. My son turned on the computer. There was an explosion, and my speakers were, for all practical purposes, vaporized. My son was very impressed with the sound effects. Was there more of this to come? Well, not from those speakers, not ever again.

Having destroyed my speakers, the best thing would have been to sit down and have a cold drink. But professional pride was at stake; the game must go on. Willing my mind to stop calculating the cost of new speakers, I thought of how I might rig up some temporary audio, and I remembered some little Walkman speakers that I had. The only problem was that these little speakers had a microphone plug connection, whereas the amplifier required bare copper wire. But any true EE has some alligator clips somewhere, and with undue haste I clipped together the old speaker wires to the microphone plug, in my hurry unknowingly shorting out the amplifier's output wires. This time when the amplifier was turned on there was no explosion at all. There was, in fact, nothing but silence. The amplifier had now burned out. Finally, I sat down to a cold drink. I told my son he would have to play the computer game another day, after I earned a lot of money to buy a new system. Never plug things together in haste.

Networks

I have spent a certain amount of time bemoaning the loss of the hobby of electronics. While I do miss those good old days of tinkering, there are compensations for living with the technology of today. You lose some things, but you gain other things. Just as I would have never dreamed that there would be a time when my soldering iron and slide rule would lie unused in a forgotten back drawer, I could not have imagined that we would have personal computers in our homes and that these computers would link us in a global community tied together by far-reaching electronic networks.

I guess fair is fair; the same VLSI that took the soldering irons from our hands gave us the personal computer. Now the deal is done, and there is no going back — technology is a one-way street of inexorable progress. Even if we could go backwards, I would not give up my computer for all the fun of tinkering. Now that I think of it, you can't take my CD player either. Would you settle for a electronic calculator?

When I am on the road, which is a lot, my laptop is my best friend. (Its battery being my worst enemy.) Basically it is only good for two things, however. One is writing, and the other is communication. The latter never ceases to amaze me. Imagine being able to reach out to the world through a portable keyboard! It is like there is this "force" among us, like Obe Kinobe describes in *Star Wars* — a greater community of engineers and scientists sharing our knowledge and wisdom.

In the few years since I wrote "The Wisdom of the Net," the growth of the Internet has been phenomenal. There are now thousands of joined networks and millions of users in about a hundred nations including, for example, Mongolia and Pitcairn Island, just to calibrate what is happening. The "few hundreds" of discussion groups in Usenet that I mentioned has grown to several thousand. I still do not know who has done all of this, but perhaps in some larger sense we all deserve the credit. Someday it may be forgotten that it all started as an academic network mostly for scientists, engineers, and whoever could be described as a computer nerd. Now it is out of our control; it has been unleashed on the world.

Before I get carried away on the general brilliance of engineers and scientists who put the Internet together, perhaps I should give the other view on this: maybe nobody made it. Maybe it just happened. Someone recently described Internet as an example of an autonomous, self-replicating mechanism. The compelling analogy is that of fractals, generating intricate patterns through layers of recursion. Users band together in local area networks, which then join in metropolitan area networks, which then join national networks, which then join global networks. Maybe the galaxy is next, who knows?

For whatever reason, Internet exists and has a life of its own. Things happen out there. Several thousand news groups buzz with the latest news, rumors, and advice about everything from the most technical of subjects through religion and philosophy. The lines hum with the transfer of files, many of which are anonymous FTP, meaning that unknown people are shopping around in computers all over the world to see what might be interesting that they can download to their own systems. The incredible aspect of this shopping is that these are not just "sales" — this stuff is free, and it represents a tremendous output of intellectual activity. A lot of people are talking about what is said to be an evolving information infrastructure. When you say things like this, people take notice. They think you know what you are talking about.

The evolution of Internet is taking shape in what is called NREN, or the National Research and Education Network. Exactly what this is going to be remains to be decided, but the original impetus came from the idea of providing gigabit (billions of bits per second) access to supercomputers. Optical fibers carry these sorts of capacities today, but the question was, what can an individual — as opposed to an aggregation of thousands of users — do with a gigabit-per-second communications line? That is the question asked in the essay "The Gigabit Network: Who Needs It?" Hopefully, you were not looking for an answer to this question, as it will not be provided.

It has been some time since this essay was written, but I still feel that the best answer to who needs a gigabit is that we do not know. Obviously, a first answer to the question is that a thousand megabit users need a gigabit. I have also heard that one radiologist needs a gigabit, but sometimes I think they just pretend to need so much. Perhaps I can relate one small personal story that sheds a great deal of light on this matter.

Some years ago I had a minor stomach complaint, and I went nervously to an unknown internist. I was placed in one of those cold, antiseptic examination rooms where I radiated fear off all the white metal walls. After an interminable period, a white-coated doctor raced into the room holding an x-ray, which he placed under the clips of the lightbox on the wall. Holding his chin in one hand, he surveyed the x-ray, oblivious to my presence. Suddenly, the doctor muttered to himself, "My God, he's had it." I shivered involuntarily.

Reconsidering, the doctor said to himself, "Wait a minute." Then he removed the x-ray, turned it left to right, and reinserted it into the lightbox. Stepping back, he said, "No, he's all right." Then seeing me for the first time, he turned to me and asked curtly, "What are you here for?" I don't remember what I said, but it could not have been very profound.

Now that I think of it in the present context, I can only ask: for this he needs a gigabit?

The answer that we do not know who needs a gigabit is not a terribly successful sell in Washington, D.C. But I do have a certain faith in Say's law, that supply creates demand. This is also known as the *Field of Dreams* approach — if we build it, they will come. I can see the gigabit-needers walking dreamily out of the corn. Whatever it is that they do will be wonderful. So I can only say, "Play ball!" On to the columns.

Rampant Technology

Recently I had to give a talk surveying progress in our field during the last decade. I got out all the old issues of *Spectrum* (and some others) and looked at the advertisements and new product notices throughout the 1970s. It made me feel (1) proud to be an electrical engineer, and, (2) scared to be an electrical engineer. It's fine for futurists to make a living out of writing about the onrush of technology, but how about us poor engineers who have to drive this runaway chariot?

A 1971 Hewlett-Packard ad shows an engineer dressed in a futuristic white suit standing at a large piece of junk identified as "The best programmable calculator system now — and in the foreseeable future." Well, apparently the foreseeable future didn't last very long, for it was only a year later that the ads began to appear for the now-venerable HP-35. Those calculators cost about $400, and I recall the security cradle chains that were required to protect them from theft. Today if I passed a calculator lying on the sidewalk, I just might not bend over to pick it up, except that it would be considered litter.

Now a certain sense of fair play prompts me to confess that HP dressed up that 1970 ad by including in the calculator photo a Bell System Picturephone. Well, the calculator didn't last, and the Picturephone didn't arrive. It's the usual story; we continually underestimate the pace of technological advance, and overestimate the social or economic acceptance of these advances. Needless to say, there are no ads like that in the journal you are now holding!

In 1970 a 64-bit random-access memory was being announced. But by 1974 a 4-kilobit RAM was listed, and before the decade was out 64kb RAMs were developed. A small note in January 1973 described a "computer on a chip" — the Intel 8008. In retrospect I thought the microprocessor should have been announced with great fanfare, say on the front page of the *New York Times,* but it just crept up on us. Little did I realize that by the end of the decade we would be inundated with those little machines. Their power had grown from the controller-like 8008 to such mainframe-like machines as the Motorola 68000. Along the way bubbles grew (and possibly burst), optical fibers were replacing copper wires, floppy and hard disks were developed, and we heard about Josephson junctions, charge-coupled devices, and video disks. If you had been a Rip Van Winkle who went to sleep in 1970 and woke in 1980 working in some electronics firm, you would have been quickly consigned to early retirement or upper management.

Now, progress may be our most important product, but has anyone inquired whether or not this much of it is really good for electrical engineers? Perhaps we should institute a study to determine if prolonged exposure to a high rate of technological progress might not be injurious to our health. After all, aren't we entitled to some job stability? What kind of working conditions must we endure when we're forced to open every journal with trepidation, lest our latest developments be rendered useless? Will it be necessary to put the following label on everything we make? "Warning! The presidential science advisor has determined that the enclosed product is obsolete." Or will each product be essentially an empty box containing only a small coupon "redeemable for the latest insides"?

When I was in school I thought that vacuum tubes, like diamonds, were forever. The world had always been composed of various combinations of tubes, resistors, coils, and capacitors. Invention was only limited to the number of permutations possible using these elements. When I first heard about transistors, I thought hopefully that they would go away, and for a while I stubbornly refused to learn about them. Now the world has so turned that each time I hear about a new device, I assume automatically that it will supplant everything I already know about. Frantically, I give up my unfinished study of the present devices in order to turn to the new one. Naturally, the continuation of this process has left me without any discernible base of expertise. Each night when I go home I nervously sift through the mail in expectation that there will be a letter from my alma mater issuing a general recall of degrees issued in certain years, among which will be my year, because of certain unspecified defects which have made them unsafe for use in the outside world.

I remember the wisdom of a classmate many years ago who decided that he wanted to specialize in 5-ohm resistors. In the years since then tubes have died and transistors have come and gone, while resistors have remained. However, even here the onslaught of progress has been inescapable, and the 5-ohm experts have been retrained or replaced with 1-kilohm experts. If not even resistors are safe, what else is there? Vainly I cast about in my mind for an occupation where an accumulated knowledge has a lasting value. For awhile in my fantasies I considered archaeology, specializing in, say, Greek vases of a certain period. But then I thought that even here I would probably have to be running computer analyses of all sorts. We're not only making things difficult for ourselves but we're complicating the lives of others too!

I myself feel blameless for the creation of all this turmoil/progress. As I look about it occurs to me that all my friends also seem innocent. Somebody else must have done it. Perhaps there are just so many of us that, without anyone actually meaning to cause progress, it happens without our conscious intervention. Sometimes I think that we could stage a small job action, agreeing not to invent anything for some specified period. However, I have a fear that we would find subsequently that some revolutionary device had been invented during the job-action period. We would all be forced to take courses or otherwise learn about this new device. Meanwhile all of our accumulated knowledge about something important like logic design would have been made worthless.

My nightmare continues that we are unable to find out who invented the new device, so we have no one to blame for this transgression. After further intensive investigation we discover that the dizzying rush of technology is really an extraterrestrial plot. I see the alien chuckling to himself as he types yet another new product advertisement for *Spectrum* on his ancient Remington upright, with soft music from an old Philco tube-type radio in the background. Or I see the aliens prowling the corridors of Intel and Texas Instruments at night, sprinkling new parts and documentation. "This will keep them confused for another three years," they laugh among themselves as they conceive of another complicated signal-processor chip. "This way they'll never catch up to us."

The Incredible Shrinking Transistor

Twenty-five years ago, on the occasion of the IEEE's 75th anniversary, all IEEE Fellows were invited to write a short essay on electrical technology as viewed from the year 2012. The year 2012 was selected because it was 50 years from the planned publication of the essays in a special issue of the *Proceedings of the IEEE* in May 1962. Apart from a certain entertainment value, they provide conclusive evidence of the inability of technologists to predict the future course of technology.

One Fellow's vision of the (1961) present as seen from 2012 has always fascinated me: "After a competitive race in the 1960s to produce the smallest units, reason had prevailed. While components were small by earlier standards, the ultimate sizes were such that costs were reasonable and servicing practicable. For example, whole receivers were the size of pound candy boxes rather than cigarette packs."

The phrase I love in this is "reason had prevailed." In fact we could argue that during the 25 years since this was written reason had not prevailed. The race to smaller and smaller circuitry has become a worldwide obsession. The "whole receiver" size didn't stop at either the candy box or the cigarette pack. Furthermore, the physicists (some of the best of whom are electrical engineers, of course) smugly assure us that the exponential decrease in transistor size can continue for another decade or so. Lucky us.

Now, I know that what's done is done. It would be awkward, to say the least, for us to tell the world that we made a mistake and that transistors really were never intended to be this small, but was the 1961 Fellow right? What is, after all, a reasonable size for transistors?

In 1961 the argument for reasonableness was based on costs and practicable servicing. Presently, I shall get to my reasons, circa 1984, for why transistors should stop shrinking. But as far as costs go, the exponential shrinkage has brought with it a concomitant dilution of cost per transistor. This is fine when you are a consumer, as occasionally even we are. But this also means a shrinkage of profit per transistor.

We engineers are forced to continue the dizzying race toward higher productivity measured in transistors designed per engineering hour, unless of course we would be willing to settle for an exponentially decreasing paycheck. The quality of engineering design will come to be judged on the basis of how large a number of transistors can be expended to do a given function — designs requiring fewer transistors being rejected. You are not naive enough, I hope, to believe that they really *needed* half a million transistors in the Hewlett-Packard microprocessor. Would OPEC put up with a 50-percent yearly inflation in the miles per gallon obtainable in internal-combustion engines? Ask the manufacturer of low-end home computers what the shrinking of transistors has meant to electronic profits. Are we to be like lemmings, or is now the time for reason to prevail?

Our 1961 Fellow complained about the practicality of servicing smaller transistors. Consider a present-day microprocessor chip. Do you see any assembly screws in the package? How are we to get inside? Imagine taking your computerized car to your local garage for servicing. Lots of luck. Or what happens when your electronic ignition fails in the middle of the boondocks? You stop and forlornly put up the hood, peering inside as if you could really do something about it. Personally, I don't trust anything that I can't get a wrench on.

Do you own a video cassette recorder, and if so, have you looked at what's inside it? If you have, you know for sure that no one — and I mean no one — is going to fix it if it's broken. I instantly put the cover back on mine and have unsuccessfully tried to rid my mind of the vision of the rabbit-warren of unmarked parts. I suppose I should consider it a miracle that it works in the first place. Repair is not intended for mortals such as I.

But reasons of cost and serviceability are bygone standards. The more important reasons for putting a stop to transistor shrinking are, I believe, esthetic. It is, for example, *unseemly* for so many teeming transistors to be huddling together for warmth on one chip. The other notion is the intangibility of what we've created. Let's consider these thoughts.

Remember the good old days when electronic circuits were assembled by real people using their own hands aided only by soldering irons? Nowadays we engineers are like Gulliver on the island of Lilliput. Fabrication facilities resemble hospitals, with gowned figures — hushed and ethereal — gliding about examining microphotographs that seem to indicate stages of illness in the circuitry. The only noise is the Niagara-like roar of the filtration machinery. You could almost imagine that some quirk of gene splicing had loosed a transistor-eating micro-organism.

What does this have to do with electrical engineering, you wonder? Forget about getting your pudgy fingers on these little critters. Perhaps the only way they could be approached would be to construct a robot whose sole function would be to replicate itself to half scale. We build the first robot and then sit back and watch it disappear into the world of micro-electronics, where we can never follow.

This brings me to my second worry. How do we know these transistors really exist? There are levels of existence, after all. Things that can be touched and seen have a tangibility. But when you have to take the word of an electron mircroscope, what kind of existence can you assume? It reminds me of the difficulty I had in high school in understanding the concept of imaginary numbers. If they didn't exist, what good were they? It was hard to trust my teachers that eventually such numbers would prove useful, but later in college I found what they could do in engineering problems. Perhaps the illusory shrinking transistors prove themselves in the same way — that is, only through the evidence of their utility. Meanwhile I'll have to take Motorola's word for it that there are 40,000 devices in the MC68000. Maybe they counted them. Or maybe they didn't.

The level of tangibility is directly related to the existence of magic. We engineers like to believe that we understand how everything works. Once I asked my wife if it bothered her that she didn't know how a TV set worked. It was simple, she replied in an offhand manner: you simply turned the knob on the front to "on." That stopped me; she was right. It was a fine example of top-down thought.

My knowledge of the workings of the set goes a number of layers deeper than my wife's, but there's always a lower layer where magic needs to be invoked. I need to trust the theoretical physicists for the principles of behavior in the microcities of the semiconductor worlds. In the future the level of invocation of magic will rise. Only a trusted few will be chained in the basements of the semiconductor houses. The rest of us will deal in high-level functionality and virtual, or logical, operations. A shroud of mist will form over the magic below.

What's Real Anymore?

Not long ago I was visiting a large university as a member of some committee or other. Instinctively choosing my area of greatest incompetence, the university officials sent me over to the materials engineering department to inquire about whatever might be happening in that unknown territory. To begin my visit I was introduced to a group of randomly chosen students. Of course they were all articulate, intelligent, enthusiastic, straight-A students with a great variety of extracurricular activities. That much I expected, based on past experiences with students randomly chosen by interested faculty members. What I did not expect, though, was that most of them would be women. Upon inquiry, I was told that almost half of the students in materials engineering were women.

"Why are there so many women over here, as compared with EE?" I asked. Before anyone could even answer, I blurted out an ill-considered defense of my inane question. "I mean, it's clean over there, ... and there isn't all this heat and machinery ..." I waved in the direction of the labs behind me. I think I was once again being defensive about the association of the word "engineer" with the hot, sweaty, soot-covered caricature of the railroad driver.

One female student looked at me with more than a hint of incredulity and disdain. "You can see and touch things here," she said. Glancing toward the nearby EE building, and barely suppressing a shiver, she added, "Nothing is real over there." Suddenly I had a vision of EE through her eyes. It was like seeing my cherished profession through Lewis Carroll's looking glass. *Nothing is real over there.* Could that be so?

My instinct was to protest, but there is a time and place for everything, and I changed the subject. She had hit a sensitive nerve; I had often pondered the reality of electrical engineering myself. Our field has always been divided into two camps — those who work with *things*, and those who don't. Some of us live in the real world of circuits and devices, while others inhabit the netherworld of mathematics and software. And the latter part is growing — reality is slipping away.

A friend who works in materials science used to start his talks with a simple chart displaying a single sentence. "Everything must be made of something," it said. In a few brief words it captured his pride in his field. What could be more fundamental than materials work? But one characteristic of engineers is that they like to argue. "Not so," I said to myself. Most of our stuff is made of nothing at all. It is made of software, of math, of conceptual thought. We live mostly in a virtual world.

When I was in high school the sensual aspects of electrical equipment appealed to me. I can remember the smell of burning insulation, the feel of warm vacuum tubes, the sight of flickering orange filaments, and the sound of humming transformers. In contrast, I had a lot of trouble with the notion of imaginary numbers. What were they good for if they didn't exist? My teachers had a hard time explaining this to me. I don't think they knew themselves.

In college the importance of that imaginary world grew, while the physical reality of electronics receded. Mathmatical models stood in for the physical world. Even in the laboratory I viewed reality only through an oscilloscope as intermediary. Sometimes I wondered if physical behavior was really indicated by the squiggly line of glowing phosphor on the scope's face. I looked at the inert circuit on the lab bench with a touch of suspicion.

Now even the scopes are disappearing from labs. Who needs them? The computer monitor simulates a scope trace, showing the performance of a simulated circuit. Layers of software intercede and protect the sensitive human eye from actual observations. Engineers work on architecture, on increasingly abstract methods for formal specification, test generation, and validation. Nothing is human. Nothing smells. There is nothing to feel.

Today the physical circuits themselves are usually composed of anonymous chips like DSPs (digital signal processors). If someone should ask you what these chips do, you would be hard pressed to answer. "Well, it depends," you would say hesitantly, hoping to end the conversation. Unhappy with this evasion, the observer might look at the few lonely chips and complain that basically there was nothing inside the box. How can this thing work with nothing inside? But it is that seeming nothingness — the program — that makes the box do whatever it does.

I do not mean to lament the plunge through the looking glass into the world of unreality. I think that is the way technology is inevitably headed. There are too many constraints in the physical world, and functionality has become too complex to manipulate in terms of physical objects. But I confess to an occasional nostalgia for things that can be seen and felt.

A few years ago I hosted a dinner for winners of certain company awards for technical achievement. In an attempt to enliven the program I asked each winner's spouse to tell us what their mate did at work. Of course, I discovered once again that the typical spouse of an engineer has no clue about what their husband or wife does for a living. On this occasion there was but a single exception. The wife of one of the physicists claimed that she knew exactly what her husband did in the research lab. I wondered how she would explain his research in vapor phase expitaxy for growth of optoelectronic wafers. "He goes off every morning and spends the day slaving over a hot oven baking cookies," she said. Close enough, I thought — there is a down-to-earth explainability about the business of making things.

I have a number of friends now engaged in research into a field called "virtual reality." No one seems to take the title as I do — as an oxymoron. Surely this is the ultimate self-deception, the one that closes the loop. People will bake virtual cookies. We will smell the burning of virtual insulation, and hear the simulated hum of virtual transformers. We could easily explain this to Alice — it is all done with mirrors.

An Engineer in the Land of Lilliput

Tiny things — we engineers specialize in them. We watch with pride from our comfortable position in the macroworld as the creatures of our devising shrink into the microscopic dimensions of virtual nothingness. We should be proud. We are the Gutenbergs of our age — having invented and perfected a printing press for electrical circuits. We are the originators of a photolithographic process that produces tiny electrical cities that can be made in high volume for miniscule cost.

Now what else can we make with our magical printing press? Recently some engineers have been running around urging that we print mechanical assemblies on silicon the same way we print electrical circuits. After all, as unfortunate as it may be, some of the world remains stubbornly mechanical. Tapes hiss, platters spin, heads seek, motors whirr, and gears chunk. There are things out there that move. Maybe they need us and our printing press.

A few years ago some engineers in my organization made a tiny gear. It was about 100 micrometers in diameter, a giant in the land of transistors, but maybe, I thought, the smallest precision mechanical part ever made by humanity. Shortly thereafter I was scheduled to give a talk to a visiting group of executives from a competitor. The little gear would be just the thing to add a touch of showbiz to an otherwise deplorably dull presentation, so I went to visit the gear-makers. I remember the look they gave me when I asked to borrow their precious tiny gear, but management has its privileges, and I was carefully handed a transparent petri dish containing a little black speck.

The next morning I walked out of my house with the inevitable stack of Vugraphs in my arms, topped precariously but confidently by the petri dish. My thoughts were off in space as my foot reached awkwardly for the last step on my porch, a step that I had already unknowingly negotiated. A slow-motion replay would have shown the pile of Vugraphs lurching, the petri dish tumbling off into space like a newly freed satellite, the dish opening in mid-flight, and the opened dish landing daintily upside down in the slowly collapsing stalks of grass. But not even the replay would have shown the tiny black speck vanishing into the vastness of the green jungle that was my lawn.

I remember getting down on my hands and knees, and parting the threads of grass, wet with heavy drops of dew and colored by the refracted rays of the rising sun. Somewhere in that tangle of turf was my little gear, but I knew with certainty that the smallest precision part made by mankind was now history.

Still, the show had to go on. Even in the midst of my obligatory talk to the visiting executives, I realized how painfully boring it was becoming. I felt compelled to tell them about the tiny gear, and swept away with my theme, I withdrew the petri dish from my briefcase and passed it around the table. Each person peered into the dish and appeared to nod in confirmation. I had no idea what they were looking at, but they all seemed impressed.

Later I felt embarrassed. I confessed to the president of my company (who had been present at that morning's talk) that the dish had been empty. I expected a lecture on corporate ethics, but instead I got a bit of philosophy. "You know," he said, "there is probably an ant in your lawn that has discovered the wheel." I immediately realized that I had done the ant population a giant favor.

I imagined being an ant out for its morning walk in my lawn. (They probably all get up early, I thought.) Suddenly, out of the sky falls a WHEEL. It would be like that moment in the movie *The Gods Must Be Crazy* when the strolling bushman encounters the mysterious Coke bottle thrown out of the passing helicopter. I could imagine going back years later to visit my lawn and seeing this new ant religion based upon the tiny gear. It would probably be mounted on a large pedestal and surrounded by supplicant ants. (If this leads to something, remember that you heard it here first.)

Undaunted by my irremediable clumsiness, the gear-makers persevered. Next came tiny gear trains, and then people at universities making tiny motors. It is all exciting exploration, but there are always other people who ask the inevitable question: "What is it good for?" Everything, it seems, must be "good for" something — an ant religion does not cut it.

Now the mind wanders, as it is wont to do. Just imagine. Tiny robots, controlled from our lofty position, that could visit the world of Lilliput — little robot slaves that could crawl around our microchips and walk where we are forever forbidden to tread. But again, why? What is there worth doing in those microcities when the cities themselves cost almost nothing to replicate?

But wait. There is one microworld worth fixing at any cost — the one inside our bodies. The biomedical people leap at the chance. How about a tiny Cuisinart for our arteries? How about — at some time in the distant future — a microsurgeon that will do the job? You just swallow a little pill, and as it dissolves in your body, out jumps a little robot with a tiny "S" on its T-shirt? Microsuperdoc! It hums quietly to itself while it hacks away at the offending tissue with its micromachete. You lie on the operating table like Gulliver, contentedly reading a book, while the superfluous human surgeons watch their computers display telemetry from the microsurgeon.

My restful fantasy takes an unexpected turn. One human surgeon raises a questioning eyebrow at his CRT display. He glances quickly across the room to an associate at mission control, who looks up in alarm. A nurse, also large and human, manipulates a joystick and nervously presses a key repeatedly. The microsurgeon has run amok! It is still humming and hacking with its micromachete somewhere deep in your body. The calm, bored voice comes on the PA system. "This is mission control. We have an abort. We have an abort." The book you have been reading does not seem to hold your interest any more. Somewhere in the background you overhear the urgent whisper, "Why doesn't it respond? Can't someone turn that thing off?"

Goodbye, Heathkit

The other day a small company in Benton Harbor, Michigan announced that it had given up on one of its lines of business, and would henceforth concentrate on other products. In the business community this event passed without notice. Had Wall Street reacted, the analysts undoubtedly would have cheered the elimination of an unprofitable division. Smart decision, they would have said, and the stock would have risen.

For us electrical engineers, however, the decision of that little company signified the passing of an era — the end of a time when it was fun and profitable to tinker with electronics. The product that symbolized the electronics hobbyist is no more. It hurts to say it, but say it we must: Goodbye, Heathkit. We shall miss you.

Some of my previous articles have touched on the theme of the disappearance of electronics as a hobby, and I have sensed from the letters I have received that this has sparked a note of wistful remembrance, at least in the older engineers. One engineer writes about hearing the news of Pearl Harbor on a crystal set that he had just finished constructing. Another laments the closing of "Radio Row" in lower Manhattan, where electronic hobbyists could wander from store to store in search of surplus parts and kits of all sorts. That space is now being trampled upon by several tall buildings known as the World Trade Center. The letter writer suggests that perhaps today a single memory chip would contain more circuits than did all of Radio Row in its prime.

I remember studying the new Heathkit catalogs, calculating my available dollars against my limitless cravings. I remember the thrill of seeing the big boxes with the Heathkit return addresses waiting for me at home. I remember the great satisfaction in sorting through the little bags and boxes of parts, and slowly beginning to create order out of chaos. (There is nothing like defeating the second law of thermodynamics.) Where there was only a tangled jumble of components, like a game of pick-up-sticks, a beautiful television or hi fi would begin to take form and grow. Check mark by check mark the progress would be recorded in those famous manuals — signposts along a yellow brick road leading in a measured pace toward the ultimate reproduction of the picture in the catalog.

As the kit neared completion, the expectancy would grow, until the penultimate moment of pregnancy and doubt arrived and all that remained was to turn on the power. The thrill of victory, the agony of defeat — it all awaited. Generally speaking, I always had confidence. After all, I was an engineer. Surely there would be no problem that could not be bested by my knowledge and experience. Unfortunately, that was not always the case. Having assembled their most expensive color television, I could only get a raster display that was broken into shards of color resembling a church window done by a modernist painter. My wife looked worried. I exuded false confidence.

More and more my thoughts were pulled toward Heathkit's motto, "We won't let you fail." But the thought of pleading for mercy and help alongside all the accountants, shopkeepers, and grade school kids who built these kits was demeaning. The advertisements all said that you did not have to know anything about electronics to build the kits, but I always told myself that surely my engineering training had to count for something. Secretly, of course, I knew that was wrong. You could build their most difficult kit and not learn a single thing about electronics. The television sets that the grade school kids built probably worked; mine didn't.

I planned my telephone conversation with the Heathkit experts. There were two strategies. First, I could come across as an expert, who just needed a little professional-to-professional consultation. I could tell them about how I had analyzed all the scope patterns, and about my diagnosis of the problem, etc. Alternatively, I could pretend to be in the lawn maintenance business or something like that. I could laugh a little, and say some really stupid things, and let them treat me like a patient in the Mayo Clinic. Even as I dialed the help-line number, I was undecided as to what strategy would save my television, while leaving my ego intact. When the expert came on the line, I stumbled over my words, saying something incomprehensible about a broken-up picture. "Send us the picture tube," said the expert gruffly. I started to explain about the scope traces, and discovered that I was talking into a dial tone. The new tube did work, although my ego never recovered.

After the big bang of kit creation, I would enjoy a prolonged, warm period of praise and self-congratulation. During this period I would assume the self-important posture of first-cause rationalization. Whatever anyone would say, I would reply, "I saved money." My wife might comment to some friends about how nice it was to have an engineer-husband who could build these incomprehensible things. I would smile knowingly, and say, "I saved money." The pizza delivery person might catch sight of the television, and say that he never heard of this brand called Heathkit, and how come it didn't sound Japanese? I would shake my head ambiguously, and say, "I saved money."

Alas, the day finally came when I would stutter, "I saved mon... mon... mon..." I mean, how could a grown person say, "Well, actually it cost me a lot more than I would have paid for a deluxe Sony, and it isn't as good, but I really like to solder all these little wires together?" I'd probably have to twitch as I said it.

Because of VLSI, or because of modern manufacturing technology, building a kit does not make sense anymore. There is nothing inside today's boxes except a few unmarked VLSI chips with 50 gazillion pinouts surface-mounted to a multilayer board. The labels all say "no user serviceable parts inside." It's funny, but I have this vestigial memory that tells me that if I drop a piece of electronics, it will break — like it is full of vacuum tubes or something. Actually, it is just a big chunk of silicon.

An engineer wrote to me about building a computer kit in 1977. "That was a major thrill when BASIC came up and announced 'READY' on the screen. Last year I purchased a new IBM clone, brought it home, and plugged it in. That was about the same level of excitement as buying a washing machine."

Now when electronics hobbyists get together, the intelligent conversation runs something like this: One hobbyist asks another, "What kind of computer do you have?" Whereupon the reply is, "I've got a 20 Mhz 386; how about you?" Now the first hobbyist smiles condescendingly, having trapped the other, and says that he has a 486 machine running Windows at 25 Mhz. He then asks about RAM and disk size. This same conversation is repeated endlessly without substantive modification. There is nothing else to talk about.

Of course, the same VLSI that took Heathkit from us gave us the personal computer, and I would not trade my computer for all the Heathkits. So there are compensating advantages for the end of tinkering, including perhaps a new accessibility of the engineering profession to those people who did not grow up as tinkerers.

Still, it must be said: I miss that big box of little parts. I miss the growing line of check marks in the yellow manuals. Bravo, Heathkit. Thanks for the memories. Thanks for a job well done.

The Electronic Hobbyist

Electronics used to be fun. Maybe it still is, but sometimes I have doubts. When I was a youngster, I discovered a book in the library entitled *Boy's First Book of Radio*. It was an old book even then, but now it would look archaic. Each chapter gave instructions about how to build an ever more complicated radio, starting with a crystal set in the first chapter and ending with a superhet in the last. I was enthralled with the adventure of it all and eagerly started to build my very own crystal set.

My first great discovery was that crystal sets did not work. I built them both from scratch, and from kits, but I never got any of them to bring in a single station. That piece of rock on the end of the cat's whisker was a bad idea — solid-state electronics was not ready for the big time. However, undiscouraged, I found that miracles could be wrested from vacuum tubes. All you had to do was to wire them up in endlessly possible configurations, and you could pull voices out of the ether. This was good stuff — electrical engineering was for me.

Transistors came along, but no matter, they were just like little tubes, and by wiring them together with resistors and capacitors, you could do neat things. Now I concentrated on kits. See my great hi-fi system? I built it myself, saved a bundle, and if anything ever goes wrong with it, I can fix it. See my TV set? Yep; built it myself. A beauty if I do say so. People marveled at how clever and dexterous I was. Not that it was very hard, following instructions like: "Connect a 3-inch hookup wire between connector KK (S-3) and QQ (S-2) of the IF-audio circuit board." It gave me a feeling of accomplishment and pride in the finished product. I have some of those kits — still in working order, too. I keep waiting for them to break so I can fix them myself.

But then something changed. Integrated circuits came along, and all those transistors and resistors got scrunched into little chips. Worse yet, all the wires were in there, too. The external wires that connected the chips together were etched onto a printed circuit board. Nobody cut that little wire and wrapped it around the solder lug anymore. They still sold kits, but now all you did was stuff the parts onto the board and solder the connections. It still felt good, but I began to wonder why I was doing this.

Just about the time most of the fun had gone, personal computers came along. Altair got all the experimenters excited. The microprocessor was a fantastic engine, but it was only a single chip. Lots of other stuff had to be designed and wired, and hardly any software existed. The field was wide open. I was more proud of my home-designed computer than of any of those hi-fi kits. This was more like it!

Alas, that lasted only briefly. Now my third-generation home computer is humming quietly to itself while I write these words on one of those ubiquitous word processors. There is something wrong with the computer, but I haven't the chance of the proverbial snowball of fixing it myself. I don't even know what is inside the case anymore. The very large-scale IC chips confined there have only cryptic markings on them. There is no circuit diagram available anywhere for this clone without a brand name, and, of course, nothing is socketed. How could a respectable engineer fall so low?

I never see ads for kits anymore. It costs more to package a kit than to build the finished product. When you see the PC boards go through the factory, you realize why it makes no sense to wire or solder things yourself. Chunk, chunk, chunk — another perfect board rolls off the line. Ever try to buy the parts that go onto that board? Forget it. They cost a lot more than the finished and tested board. And so what if something goes wrong with that board where nothing seems removable? Buy a new board. Big deal. Chunk, chunk, chunk.

For a while, software seemed the salvation of the hobbyist. The hardware industry had standardized everything anyway. Even if you wanted to design your own system, it made no sense. Only one or two designs were supported the world over. But in software there was infinite variety. Everyone could do his own thing. I wrote operating systems, compilers, editors, neat programs. It was fun, and it was educational. Good for me.

Before I realized its transience, the golden age of personal computers ended. One day I looked around for some program to write, but there was nothing left. Anything I could think of had already been packaged as a commercial program that worked far better than one I could ever write.

Worse yet, there was always a free program that was better than anything I could accomplish. There was no excuse for building either hardware or software. You couldn't save money, and you couldn't make anything different.

The magazines that used to have circuit diagrams and software code just turned to reviews of commercial products. Last week, I went to a computer flea market that has been a regular source of experimenter junk for over a decade. Two discouraged men passed by. One shook his head sadly and said, "It's all gone commercial." I raised my eyes and surveyed the field, and I realized that I was looking at five hundred stands and booths all selling the same two dozen commercial products. What was I doing there?

Now what? How about *Boy's and Girl's First Book of VLSI Design*? The kit comes with a bunch of CAD software disks and a certificate to send your finished design in to the VLSI foundry shuttle for fabrication. Or what about *Build Your Own Molecular Beam Epitaxy Machine in Your Spare Time*? And with all the research on finer linewidth using X-ray lithography, maybe there will be a market for home synchrotrons. Be the first to get your neighbors together and run the loop around your block.

I hear that freshman enrollment in electrical engineering has been dropping steadily since those halcyon early days of personal computing. I'm looking at my nondistinctive, keep-your-hands-off clone, and I'm wondering — do you think there is any connection?

The Paper Airplane

Not long ago I sat in a meeting listening to a seemingly interminable talk on what I felt was a very marginal project. The speaker was a member of an aerospace company that clearly lacked for no facility in the production of photographic transparencies. I stared at the dozing audience, wondering for the nth time how we can be so placid in such trying circumstances. Not a movement or spark of life betrayed the origin of the paper airplane that was slowly circling over my head. With a small whish of air it landed at my feet, without attracting the slightest notice from my catatonic neighbors.

Inside the folded airplane was a scribbled confession. "I have just become a convert to Gramm-Rudman," it said. Beautiful! The Gramm-Rudman act is presumably designed to balance the nation's budget by putting financial pressure on just such excesses as I was witnessing. It isn't a thought that occurs often in my mind, but the message in the airplane made me worry anew about who pays, and why, for work such as ours.

I am not so concerned about those of us EEs who deal directly in the design, sale, or manufacture of a product. In that case there seems to be a direct relationship with a market. But many of us have less tangible proof of our value to society. Funds descend from the nether reaches of government agencies or corporate coffers. Studies are conducted, equations are manipulated, and possibly experiments are implemented. A final report is written and sent off to the sponsor. Nothing is ever heard about the report again, but more money is forthcoming for yet another study.

Recently the U.S. Department of Defense was chastised for buying ashtrays that cost $600. Senator William Proxmire gives the Golden Fleece award to agencies that sponsor studies on such subjects as the mating habits of left-footed alligators. But the public gives little attention to a technical-sounding study in which it is all but impossible for a layman to understand the title. Such studies are by definition good. (Lucky for us.)

I have no doubts about the societal value of electrical engineering. Collectively, we have done great things, and I feel proud to belong to our profession. It is just when I get down to a given individual contribution that I have qualms of conscience. Is a paper entitled "A Functional Analysis Relating Delay Variation and Intersymbol Interference" worth the considerable sum someone spent on it? (Best we don't deal with actual figures.) I can fearlessly pick an arbitrary title like that one, since I wrote that paper myself many years ago. I think I can safely say that it has passed the test of time. It has proven utterly useless over a considerable number of years.

Nevertheless, I can always invoke without a blush our well-traveled catechism that one never knows with certainty when a given contribution will become appreciated and important. This saves us from the need for embarrassing real-time value judgements.

I am often reminded of the famous lithograph by M. C. Escher that depicts a pair of hands drawing themselves, half in and half out of the plane of the paper. So it seems sometimes with our technical literature — both creating itself and begetting itself.

Of course, we have a system for judging the value of manuscripts or proposals. It is called a committee of peers. In other words, us. It's a pity there isn't anything better, but that's it. No one else can understand this stuff. Let's face facts — we have enough difficulty ourselves.

Somehow I'd prefer something a little bigger than ourselves — something like "consumer reports" for engineering studies. "The following papers are rated unacceptable," it would say. One hopes one's own paper would be a "best buy." But such is the stuff of dreams. The burden weighs heavily upon us alone, and often now I recall how useless I felt Reed-Solomon codes were when I first read that abstract theory many years ago. Now millions of these decoders are going into our homes in the innards of compact disk players. In our homes, of all places, and for only a few dollars! What has the world come to?

But one such latent, belated success doesn't absolve us of the responsibility for applying a certain test of probable worth to the proposals and publication submissions that are the common currency of our trade. I know that much is easily said, but like you, I'm sure, I've had a difficult time through the years telling what is good and what isn't.

I have this terrible fantasy of being thrown out of this kind world into one in which there is little appreciation for abstract thought, and hence no general system for centralized funding of research. I imagine setting up my roadside stand, selling communication theory by the page. Every now and then a passing motorist stops for the free lemonade I offer and looks at the samples that adorn the walls of my stand. "What is this stuff good for?" he asks.

I patiently explain that the paper he is examining will predict for him the anti-jam margin of a certain form of spread spectrum modulation. The particular system is rather detailed and a little unusual, since otherwise similar papers would have appeared in other roadside stands. Fortunately for him, this particular paper is on sale this week. I see the motorist nervously eyeing his car and placing his half-finished lemonade on the counter as he mutters something about already having one of those at home.

It is wonderful to live in a world rich and sophisticated enough to appreciate the value — in the large — of engineering and scientific studies. The case for accepting such support for honest endeavor without qualms about societal value was argued to me by a friend who had learned a lesson in such values during a brief midlife career crisis.

Eschewing the theoretical work that had earned him great honors, he quit his corporate research position to join a government agency where he might devote his energies to solving important societal problems. Three weeks later he returned to his corporate research job, chastened and silent about his experience, but with a new appreciation and acceptance of his role in technology. Although he never spoke of the government agency, an associate explained to me by recounting the following parable. It seems that a man had spent World War II in the armed services in charge of an "LMD." (I determined that this acronym referred to "Large Mahogany Desk.") When his children would from time to time ask him what he did during the war, he always replied, "It was an experience too horrible to relate."

Thus I suppose the value of our research is not easily weighed by measuring the worth of individual papers or reports. It is only the nurturing of the underlying system that counts. I am explaining this to my paper airplane souvenir. It doesn't appear to be listening, and its message remains. "I have just become a convert to Gramm-Rudman."

A (Very Modern) Connecticut Yankee in King Arthur's Court

Mark Twain's famous novel about the nineteenth-century engineer who is mysteriously transported back to the Middle Ages of King Arthur has always fascinated me. Imagine the opportunities in being the only engineer in that world of chivalry and magic! Of course we have come a long way since the telegraph of Twain's time, so a contemporary version of this fable for electrical engineers would have to incorporate more seemly miracles in the spirit of modern microelectronics.

Picture yourself there, suddenly cast by some cataclysmic event into the Middle Ages, complete with your twentieth-century EE degree. You never learned swordplay, and if you are like me you can't even handle a horse with much aplomb. But you alone in all the world know Maxwell's equations! (Or at least you think you do.) How to gain an advantage? Or perhaps we should merely ask, how do you survive?

At the gate to the castle you proclaim yourself to be the gifted white knight from the West, potential worker of great magic and general seer. Although your armor seems tarnished (a bug in the cataclysmic event), your strange twentieth-century accent carries the day, and you are admitted to the legendary Round Table — temporarily, of course, pending your production of a suitable miracle. You consider your situation. The king is after you to come up with a schedule and list of milestones for your miracle. Merlin seems jealous and is preoccupied with concocting spells around a small model of a white knight with tarnished armor.

Guinevere is casting flirtatious glances (but you know better than that!), and you are being constantly shadowed by Mordred. The pressure is on.

Each morning most of the knights ride forth to pursue their arduous daily labor of searching for the Holy Grail. Occasionally, for a break in the monotony, there is a war. Opportunities for electronics are boundless. Not wishing to rush headlong into the production of unappreciated miracles, you do a careful analysis of market segments, passing out questionnaires you have had drawn by the illuminated manuscript department of the nearby monastery. Finally, reluctantly rejecting the possibility of manufacturing metal detectors for grail searching, you decide to enter the personal-computer business. A large untapped market awaits your entry.

Being the first in the field, you have a wide open choice of names. An apple with a bite removed has unfortunate connotations for this society, you believe, casting your thoughts toward Guinevere. (My women engineer friends may substitute Lancelot.) On the other hand the name "IBM" has no market appeal. Texas has yet to be named, and Japan is unknown. "Radio Castle" would be OK, but it would mean you would have to go to the extra difficulty of inventing the radio. By the process of elimination, you found the Knightsbridge Computer Fiefdom.

Next is the minor matter of funding. You explain the advantages of a personal computer to King Arthur, pointing out the considerations vis-a-vis a mainframe for the castle and assuring him that he would possess a beta test-site location. The required funding of 17.3 GNP will begin to pay returns in just a little over a century, according to your discounted cash-flow analysis. You tactfully suggest that from your perspective in the twentieth century, these GNP were largely wasted anyway, unless he cares to invest in a large cathedral, in which case you will introduce him to a friend in the real-estate business.

King Arthur becomes convinced when you describe how visi-siege software will enable him to play "what-ifs" on castle inventory status. The kingdom payroll information can be kept on dScroll2, and all the knights will be crazy about adventure games. To clinch the deal you also hint darkly about things you have read in the future about goings-on in the castle.

The spoils of a recent victory have put a large supply of very cheap labor at your disposal. You smile as you realize that this will mean you will not need to seek offshore assembly capability. Now the fun begins. Eschewing with reluctance the IEEE 696 standard, you decide upon a modified (because you have forgotten portions of the circuitry) IBM PC design, reasoning that this way your computer will be compatible for the big market by the time the twentieth century rolls around.

Small adjustments of the castle economy are now required. The vase-designing artisans in the basement are converted into a CRT-manufacturing subsidiary. Later you will invent home television to bolster sales, you remind yourself. The monks will make the beginnings of a semiconductor arm, which you dub Turret Instruments. You have visions of a fat yellow "TI" catalog of ICs in the near future. Once electronic warfare catches on you will be bound to make a profit in this line of business alone. Fortunately, the monks are already dressed for the part, and those who have been attempting to inscribe the Lord's Prayer on the head of a pin are put to work in clean rooms on the initial fabrication of the 8088 microprocessor that you require. Knights who insist on visiting the fabrication facility will have to have their armor grounded.

By this time you are a big person on the Camelot campus. Lancelot is miffed and has challenged you to a joust. You have one week to acquire the skill to face the bravest, purest knight history has known. The oddsmaker Jimmy-the-Briton has you on the short end of a 10,000-to-1 wager. Mordred has started a takeover bid on your company. Guinevere is becoming openly seductive, and King Arthur wants to move up the delivery date on your computers. Meanwhile Merlin is rumored to be preparing to enter the low-cost home-computer market and is audaciously planning to name the product after himself. You yourself would like to be software-compatible, but with an eye-of-newt processor, it looks difficult. A miracle of some sort is required quickly. Back to Maxwell's equations.

Could you do it? Alone in a medieval world, given cooperation and unlimited resources, could you recreate a modern computer? Or are there glaring deficiencies in your knowledge? Could you even make a single transistor? Mark Twain's hero encompassed all of technology, for practical purposes, in his capabilities. We have come so far since then, and our field is so large and complex, that I think it unlikely that a single person could fabricate a computer in his lifetime — we depend too much on each other's knowledge and on the framework of modern support technology that we have built. But it's an interesting mental experiment. Try it.

Coping with Complexity

Is there anyone among us who has not at some time yearned for a simpler world? The daily deluge of technical literature reminds me that although I have taken a vacation, technology has not. Someone has been working, and technological knowledge continues to expand at what I presume is an exponential pace. All about me is the evidence of evolving complexity. Perhaps the second law of thermodynamics is responsible. Entropy — disorder — is steadily increasing.

I think fondly of Walden Pond or Shangri-la, but in truth these are no places for electrical engineers; we make our living from complexity. If things were simple, there would be no need for our skills. Fortunately, our very work helps create the unfolding web of complexity that will ensure our future livelihood — if only we can keep our professional sanity along the way.

When transistors were first used in products, we carried over the notion of complexity as learned in the world of vacuum tubes: the complexity of a circuit was proportional to the number of transistors. Radios were advertised on the basis of this number. I remember having a slick portable with the number "7" emblazoned on its front, indicative of its clear superiority over the lesser models with only 6 transistors. Today, a portable radio might be labeled "127,615," except that such a label would obviously be absurd. Who cares whether a product has 100,000 or 200,000 devices? As engineers we believe in achieving the most economical design to accomplish a given function. A large portion of the art of engineering lies in having a feel for economy in design. The number of individual devices is no longer indicative of economy.

To the hardware designer, the number of IC packages — or for a larger system, the number of boards — serves as the most useful measure of complexity, in spite of the degree of complexity within the individual integrated circuit. To the hardware designer, each circuit is a little bundle of complexity, tied up with string and bearing a little notice that says, "This side up. Fragile. Do not open." This is the typical approach to conquering complexity. We build hierarchies of elements that hide the inherent complexities of the underlying elements. After a while we forget that we have built the hierarchical structure and that there are elements of a different nature underneath. When a rock is turned over, strange, disagreeable things crawl out; we learn not to turn over rocks we have constructed.

The hardware designer congratulates himself or herself on the completion of a design, thinking all that remains to be done is the software. We've learned better now, but for years we underrated the complexity of software. For the most part, the hardware designer, through microprocessor and memory architectures, has merely swept the complexity under the rug of software. For most current computer systems, the overwhelming contributor to complexity is the software.

About half the programs written in my company require more than a million lines of code. Think of the complexity of a million-line program! Surely the generation and maintenance of such programs is beyond the capabilities of us mortals. And yet, undaunted by the nightmare complexity of such undertakings, we now develop million-line programs — if not routinely, at least according to schedule and budget. We obviously have a long way to go to cope with the complexity of software.

Hardware designers point with pride to the beribboned little packages that they have created to handle complexity, and they taunt the software specialists with being unable to produce the equivalent. I think the hardware task is much the easier, since hardware designers deal in fossilized software and their medium is reinforced concrete, not easily changed by any passerby. Furthermore, the number of pins on their packages constrains the input variables. Nonetheless, there probably is something to be learned from the hardware designers; certainly, the issue of reusability is pertinent for software.

In the last decade, a remarkable mathematical philosophy dealing with the complexity of computation has emerged. We have discovered a class of problems, called NP-complete, which are thought — for all practical purposes, *known* — to be intractable. That is, for a problem of this class, as the number of input variables grows, the amount of computation required increases exponentially. To put it simply: don't plan on being able to solve large problems of this type. The familiar traveling salesperson's problem of minimizing the distance traveled in visiting a given set of cities is a classic example of an NP-complete problem. Mathematicians and engineers have accumulated in recent years a long list of problems known to be NP-complete. The discouraging part, it seems to me, is that every problem of interest these days turns out to be in the intractable class. Scheduling problems, testing, communications protocol verification, correctness of programs, LSI layout optimization — you name it, it's intractable.

In a sense, intractablity relieves the engineer of the burden of optimality: the problem is so certifiably complex that there is no optimum approach. Nowadays, many speeches begin by proving intractability; then the speaker can be justified in the subsequent ad hoc, unesthetic, and sub-optimum approach. Is it comforting to know that the complexity around us is so deep-seated that there are no optimum solutions to most of our everyday engineering problems? The real world certainly belies all the textbook-problem training of our youth, since in reality the answers are never to be found in the back of the book. But can we live with the fact that the answers cannot be found anywhere?

While we have been busy creating this complexity about us, and while the IEEE has grown from one or two groups to about three dozen societies and countless publications, and while the accumulation of knowledge in our field has grown exponentially, engineers are still graduated after four years of college and are often just as capable as their more experienced compatriots. Are the students smarter? Is the most recent portion of the exponentially growing knowledge base the only part that counts? Or are engineers trained only by learning the methodology of accessing and using whatever knowledge base is extant?

Sometimes I even wonder exactly what an electrical engineer is. One thing I know for sure — he or she must cope with complexity; it is the essence of our profession.

Software Productivity

The productivity of software professionals has been estimated at about 700 lines of code per person per year. Unfortunately, we seem to need a lot of lines of code these days, so that means a lot of programmers and a big budget. Programmers ought to do better, people say. That way software would be cheaper, they argue. So all we have to do is to increase software productivity, and our troubles will be over. What a wonderful idea!

Somehow a meager 700 lines per year conjures up the image of a slothful profession. We imagine the programmer up and at his terminal at the crack of dawn. Before the sun has reached high noon he has typed, "if (test = = TRUE) break." Speaking of "break," that was a tough morning. Time for lunch. After a backbreaking afternoon typing "else return(c)," he staggers home exhausted.

I am reminded of the story of the late M. Thumps, related in Stan Kelly-Bootle's *The Devil's DP Dictionary* (under "cursor"), whose atrophied corpse was discovered slumped over his remote terminal two years after his system had been shut down. There is said to be a museum-shrine to Thumps in London, where "by inserting a coin of modest denomination into an adjacent slot, the pilgrim is rewarded with a macabre reenactment of Thumps's final attempts to log on."

It is a good thing that there is such valiant energy on our side of the terminal, because on the other side the bulldozers are busy pushing up the lines of code that need to be written. A NASA official said that 100 million lines of code would be required for the new space shuttle, for instance. On first hearing this gloomy prediction an ingenius idea occurred to me, which I now donate to the common cause. If the rocket can jettison its propulsion system, why not do the same with software?

I envision the moment that the solid fuel booster is separated. Immediately afterward the shuttle shucks off 40 million lines of code into space. I see the giant rocket surging upwards with its newfound freedom. When I shared this idea with an associate, he did me one better. "Why not use the code for propulsion?" he suggested. "Simply throw it out the back of the rocket line by line to generate forward thrust!" (This is how great new ideas are born.)

Of course there are those detractors who say the voracious need for software is a hoax perpetrated by engineers. In his outrageously perceptive book, *Augustine's Laws,* Norm Augustine observes that at the current rate of introduction of electronics into airplanes, electronics will account for the entire weight of the plane by the year 2015. Thereafter the avionics industry will be faced with a dilemma: its continued profitability will require something that weighs nothing, occupies no space, and is expensive to produce. Fortunately, Augustine observes, such a substance has been found. It is called software.

Granted that we have a problem, what is to be done? History offers some possible alternatives. Certainly the ancient Egyptians would never have put up with this slothfulness. Programmers would have been chained to their coding slabs, their chisels kept in constant motion. Perhaps the Romans would have added a drummer to keep the mandatory cadence of papyrus lines. Crises and software busts would have been resolved with alacrity. "Ramming speed!" the software driver would cry as the drumbeat accelerated.

The techniques of the industrial revolution might also be effective. Henry Ford would have set up an assembly line. As the conveyor belt passed by, each programmer would be required to bolt his next line of code onto the structure. I can see an analogy with the cake assembly line in the famous Charlie Chaplin movie. Suppose you are late with a line of code. There goes the assembly by, and you haven't got a line to add. Now you need two. Quick! There it goes again!

Modern management psychology would follow a different approach. Proper incentives could be introduced, with each programmer being paid by the line of code. Personally, I guarantee that this would increase the production of lines of code. Or perhaps we should form quality circles to discuss our programming. "How do you feel about your lines of code?" programmers could ask each other sympathetically. Students of management also worry about the Japanese production capability — visions of Honda-like code being produced in darkened factories at night by robots to the accompaniment only of a faint businesslike hum.

While it is true that the average programmer produces only the paltry amount of code to which we alluded, it is also well known that there are superprogrammers, who upon slipping into a telephone booth turn out reams of great stuff. One such superprogrammer, when asked how a particular project could be accelerated, stated matter-of-factly that it was only necessary to fire 90% of the present programming staff. Needless to say, this scared everyone, since there was a slight possibility that it would work. On a level of compromise someone suggested that perhaps 10% wouldn't be a bad idea, if only there were somewhere to transfer this retarded 10%. Whereupon another manager created the brilliant organizational maneuver known as a "turkey farm." Volunteers anyone?

Yet another solution to the problem has been whispered about in dark alleyways. Suppose, just suppose for the moment, that we didn't always write *new* code for each application. What if we were to *reuse* old code? But, we object, aren't our applications completely unique? Besides, where would be the fun? Nobody wants this to be drudgery; if they want to reuse old code, they will just have to get along without programmers. So there.

Fortunately, current software science has found the ultimate answer to software productivity. Artificial intelligence will take over, the other kind having failed. Expert systems and automatic programming will write our code effortlessly and cheerfully. Such predictions fill me with awe. The machines will know what is good for us. Software will beget software. The symmetry of it pleases me immensely. It will never work.

Our profession's reigning software philosopher, Fred Brooks, has likened the problem of increasing software productivity to that of slaying the ancient, supernatural enemy, the werewolf. Werewolves terrify, says Brooks, because they transform unexpectedly from the familiar into horrors. So it is with software. We will need a silver bullet, he says, and then goes on to argue that we have no silver bullets, and that we are unlikely to discover any. In his argument he states that "software entities are more complex for their size than perhaps any other human construct" and that "many of the classical problems of developing software products derive from this *essential* complexity and its nonlinear increase with size."

Such insightful philosophy intrigues, depresses, and exhausts me. Anyway, for the moment it doesn't matter. I just wrote my two lines of code, so I can go home now. Bye.

What's Bugging Us?

In the current popular movie *Indiana Jones and the Temple of Doom* the hero is trapped in an underground prison, about to be crushed by a descending ceiling studded with spikes. He can only be saved by his girlfriend releasing a lever that is concealed in a niche filled with thousands of hideous bugs. In a stomach-turning scene she reaches for the lever .... It reminded me of an average day in the life of an electrical engineer. Don the exterminator gear and head for the office — the bugs are there waiting every time.

The problem is that we — unlike Indiana's girlfriend — never get any respect for the zapping of bugs. There is no aspect of our work that is more time-consuming, more tedious, more frustrating, and less appreciated than debugging. It is emotionally and physically draining, you cannot possibly explain it to your spouse, and not even other engineers are the least bit interested. To top it all off, the person who cares least is your boss. Surely Hamlet could have spoken for the unknown electrical engineer when he bemoaned the slings and arrows of outrageous fortune. I imagine him wheeling out an oscilloscope to "take arms against a sea of troubles."

About a dozen years ago, I put together a tic-tac-toe-playing machine as a Christmas gift for my young son. (This was before Atari had initiated its *reductio ad absurdum* for such devices.) I finished in what I thought was sufficient time on Christmas Eve, but an all-night debugging vigil failed to get any semblance of correct behavior. My son looked at the empty space under the Christmas tree and said, "Daddy, you should have *bought* something." He will make a great manager one of these days. Despite that lapse, I have generally rid myself of any vestige of hope that a circuit or a piece of code will work the first time. The most I wish for is something less than instant detonation.

Yet as a manager how quickly I forget. A few years ago I witnessed a demonstration of a rather complex signal processor that one of my subordinates had spent months debugging. He exuded pride until I unthinkingly asked if all the bugs had been found. He gave me a withering look of the sort reserved for absolute idiots. Enunciating each word he said, "Bob, there is no last bug." Indeed, the bugs will triumph in the end, eating all circuits and code until only dust remains.

The time has come to build an entomology of electrical engineering. We must make a science of the classification and study of bugs, and we must elevate the recognition of those who venture into the woods with butterfly nets and specimen boxes. Bugs should be brought back, alive or dead, dissected, displayed, and glorified where appropriate. There should be a publication devoted to unusual bugs, and a museum for the display of rare bugs. I can imagine the glass cases containing circuits and printouts. Small, dust-covered plaques would say things like "Genus Kowalski — indigenous to the temperate zones of the northeast United States; first discovered on a Ph.D. safari in the jungles of the MIT EE department." A large photo of Kowalski, looking triumphant and disheveled as he points to a telltale glitch in the logic analyzer display, dominates the background.

There is much to be done, but a few facts are obvious. At the top level there is a dichotomy between bugs that live off hardware and those that are codiferous — that is, that eat code. Different strains have evolved in each case for bugs that infest things that have yet to work, and those that prefer to work on things that used to work. Beyond these broad categories there is a structure that is constantly in evolution. After all, bugs have been hard-pressed to stay on top of changing technology. It is a little-known fact that the first microprocessors were free of bugs. Very quickly, however, mutant bugs were produced that specialized in microcomputer circuitry. Today's bugs are highly complex and specialized.

Recent findings have shown bugs that are adaptive and intelligent. The study of such bugs has been hampered by the probability that these bugs themselves have bugs, producing erratic buglike behavior. If this is indeed true, it provides the first demonstration that Murphy's Law is recursive, that it can invoke itself.

How bugs are communicated from system to system remains a mystery. It is recognized that it is inadvisable to design systems in the humid night air of a swamp. Circuits should not be constructed in a laboratory where nearby circuits are sick. It is inviting disaster to write software on a multiuser system on which reside bug-infested files. These are obvious precautions, but controlled experimentation must be done. I would suggest that the next space shuttle incorporate a trial in which an engineer designs a circuit while tethered in space beyond the shuttle, carefully avoiding proximity to the buggy equipment in the vehicle itself. The presence or absence of bugs in the design would tell us much about the theories of contagion. It is possible that future historians will recognize the plague of black death that swept through the world's code in the 1970s and 1980s and that they will point out our current ignorance.

Through the years, bugs have battled fiercely to protect their realm. No sooner was the logic analyzer invented than a strain of analysis-resistant bugs appeared. Symbolic debuggers enjoyed a brief period of power until a high-level software bug evolved. Make no mistake about it, today's bug is sophisticated. Put in this perspective, the dreams of current engineers about fault-tolerant designs are seen to be unrealistic. Does anyone doubt that a bug will evolve that eats fault-tolerant systems?

Should fault-tolerant approaches provide a transient panacea, the world would be a duller place. Like the plight of the troubleshooting engineer, the fault-tolerant system would be put in a situation where it would be destined to be unappreciated. How do we know what is going on inside? How many dragons has the architecture met and slain? Perhaps there should be a scorecard display on the front. Also, to me the thought that there are faults in the system — but don't worry because they're being tolerated — is unesthetic. I would almost recommend an opposite course — that of spectacular fault intolerance. At the first sign of failure the piece of equipment should fall over on the floor and emit smoke. Failed ICs would roll out of their sockets and throw their legs in the air. This seems only proper; the problem now is that these ICs pretend cleverly to be alive, when they are actually certifiably dead.

In spite of the progress in device reliability, I sometimes have the perception that the bugs are winning. Remember the old days of vacuum-tube television sets? I think they used to fail more often, but I recall how easy they were to fix. You used to pull out all the tubes in reach and take them down to the local store's tube tester, where you got in line behind children, lawyers, grandmothers, and so on, who also had their little paper bags of tubes. Today when the television set fails you need a specialist. Perhaps a heart-lung machine will be needed, or the specialist will need a battery of expensive diagnostic tests, such as CAT scans. In the end you find it requires expensive surgery, and you are never covered.

Obviously bugs are here to stay. The best that we can do in the battle with this ancient enemy is to raise the prestige of the exterminator. Whenever I kill a bug myself I imagine the kind of denouement that fictional detectives like Hercule Poirot always have. There is a drawing room full of people who have failed to isolate the bug. They listen tensely to my narrative of the succession of clues that led to my brilliant analysis. Dream on, naive engineer! The only solace I can offer is the observation that one of the great thrills of our profession is the personal triumph over the accumulated adversities of nature. Nothing makes us feel so good as to make something work, in spite of the clever and ubiquitous bugs. Enjoy your sweet triumph — it will have to do.

Plugging In

I have a brand new workstation sitting here on my desk, just waiting to be plugged into communication with the electronic world that hums in the walls behind me. There is a bright metal plug on the back of my new computer intended for just this purpose. Emanating from the wall is a blue cable with its own plug for communications, an artifact of a previous computer system, now lost in last year's antiquity. Can I push these plugs together? What are my chances that it will work?

A lifetime of experience has made me cynical about plugging anything into anything else. The world just doesn't work that way. One of my earliest memories is of wrapping a cord around a metal kitchen knife and plugging it into the wall socket. Maybe I thought I was an infant Ben Franklin calling electricity down from the sky. In any event the effect was the same. There is a PROM section of my brain where forevermore the memory of that shock is, literally, burned into storage. Don't plug things together is the advice stored there.

In college we had to take a course in heavy electric machinery. Grisly stories were passed from generation to generation of students about misconnected series motors destroying themselves in grenade-like fashion, taking out teams of engineers screaming their way to oblivion. After the experiments were wired someone in the group had to plug the machinery into the main patchboard.

I still see that board in my nightmares. The connecting wires resembled jump start battery cables, tipped by chalk-sized copper plugs set in wooden dowels. It was clear that those cables were meant to carry enough amps to heat a small city. It took a certain courage and faith in the experimental wiring to plug them into the patchboard — courage that was hard to find in myself after two plugs had vaporized in my hand, to the accompaniment of thunderclaps. We would draw straws to see who had to approach the patchboard, but I had learned: don't plug things together.

In my first real job as an electrical engineer I worked with circuit design of power transistor circuits. Such transistors cost $80 apiece then, and they were hard to get. After wiring a circuit I would be afraid to plug it in. When I did, a lot of the time there would be a "pffft" — a familiar sound to most engineers — with a small but distinctive release of smoke. By now at least I was dealing with less life-threatening voltages and currents. My first reaction would be to look around to see if anyone else had noticed my blunder. Then I would fiddle with the circuit as if it were still alive and I was only readying it for application of power — just like a golfer who misses the ball and immediately takes practice swings, so no one can be really sure whether or not he intended to hit the ball the first time.

When I traveled abroad, I found to my chagrin that all power plugs were not alike. Now I have a little travel kit with all manner of adapter plugs. The adapter for England looks especially ominous. I've had enough trouble with 120 volts that the thought of getting zapped with 230 or so volts makes me wary. When I confessed this secret fear to a distinguished, elderly professor of electrical engineering in England, he looked at me with scorn. "The trouble with you young electrical engineers," he said, "is that you don't even know how to take a shock anymore." There, I must admit, he had me. I felt distinctly inferior as an engineer — a veritable rookie confronting a grizzled veteran.

But there were more complicated plugs awaiting me. The telephone plug looks simple enough, but in the late 1960s there was a clamor to allow "foreign" connections to the telephone network. People debated about what signals could be squirted into that plug without harming the network. It wasn't very simple then, and perhaps we still don't know. (It is just that no one seems to care about that anymore.)

Someone conceived of a data coupler analogous to a fuse box to protect the network against harmful signals, but unfortunately it cost a lot of money. Little did I expect that years later I would stand in a Federal court discussing these widgets in a case destined to dismantle the Bell System. "Isn't it just like an engineer," Judge Greene asked me, "to want to build a box for protection, instead of relying on the law?" I forget what I said. I know it's in the transcript, but I don't want to read it. I still think of that question, though, in connection with satellite scramblers, copy protection in software, and other such troublesome issues.

These were all simple plugs. The world is filled with plugs of such complexity that there isn't the slightest probability of "plug-to-plug compatibility." I have the same problem with plugging in my own software that I used to have with those old transistor circuits. I hear a virtual "pffft" in my mind as the system goes to sleep upon invocation of my program. I imagine the systems administrator or superuser watching as the feeble program crashes from his (or her) lofty position in the surveillance sky. Even when the computer is my own PC, I have this little worry that the machine itself will think less of me because of my ineptitude. When it comes to plugging someone else's software into mine, I know I haven't a prayer. Things just don't plug together.

Now of all the communications plugs the physical link, the RS232 connector, is the simplest. I think it was discovered in the ruins at Pompeii — the Romans probably used it to interface with their data runners. It specifies voltages (plus and minus) which are prehistoric. It even predates TTL. But it is all there is. When forced back to bare copper and elemental, ASCII transmission, RS232 is always there, and we should be thankful, because in the intricacies of the higher layers of communications protocol lies a Tower of Babel — virtual plugs misted in indescribable complexity. Unfortunately, even the lowly RS232 connector can have different configurations. So I don't think this plug I have is going to work.

(P.S. It didn't.)

The Wisdom of the Net

Has your computer spoken to mine lately? Mine seems to have a lot of friends out there. All day and all night long I imagine that I hear the steady buzzing and whispering as messages are passed back and forth over Usenet. Although the messages are intended for anyone, I am a little embarrassed to admit that sometimes I browse among them. At their best these evolving computer networks represent a new social phenomenon, full of information, wit, and wisdom. At their worst they show the dark side of our associates — the all-too-human intolerances that we sometimes feel our profession has outgrown. Somewhere in the middle, between the wisdom and the sludge, is a growing pollution of junk bits, threatening to cover the surface of the earth with used disks full of unread messages.

In case you haven't partaken, the rules of networking are simple. There are several hundred subjects available on the net —*net.consumer* to exchange consumer information, *net.philosophy* to discuss philosophy, *net.movies* for general movie gossip, and *net.movies.sw* to discuss the *Star Wars* saga, to name a few examples. Anyone who is reached by the tentacles of Internet can insert or read news under any category. These tentacles reach far and wide in mysterious ways across several continents. No one seems to run the net; as far as I can tell, it just grew itself, and is administered by the collective will of all who tinker with it. I was surprised to find that my own computer was acting as a main hub to relay messages to Europe. Presumably it assumed this role on its own because when all the old telephone bills came to light, no human being seemed culpable.

People on the net use their own names, but sometimes I get suspicious about their affiliations. The society of "Chartered Accountants Who Want To Be Lion Tamers" sounded phony to me. On the other hand, many users with real affiliations take special precautions to distance themselves from their employers. Witness the following sign-offs:

"I suspect that [name of university] would deny knowing me, let alone share my views."

"The opinions expressed herein are randomly generated by my keyboard, and probably do not reflect the opinions of the university ... which are also randomly generated by my keyboard."

"This may not reflect my opinion, let alone anyone else's."

"The [government agency] is paying me to post this message, but if I am caught they will disavow all knowledge of my actions."

Others insert a tag line with a bit of philosophy, such as:

"It's the thought, if any, that counts."

"I've had a great many troubles in my time, and most of them never happened."

"I'm an engineer; I don't have to read."

"If things don't change, they will probably remain the same."

Usenet is best at collecting opinions and information about specific technical matters. The request "Does anyone out there have any advice about graphic boards for an IBM PC?" will never fail to draw generous responses full of useful information. The net is at its most interesting when a particular topic goes supernova and generates hundreds of opinions before it burns out. For example, in *net.movies* the question debated was "What was the worst movie of all time?" The most popular answer seemed to be "Attack of the Killer Tomatoes." But the net is far from representing an average slice of the population. Walter Mondale won the network poll for the U.S. presidential election handily. The movie *Somewhere in Time* got (I thought) terrible reviews in the press, but netters were fascinated by the concept of time travel and were passionate in their defense of the movie. There was one exception, though, from a respondent who, after reading all the praise on the net, had raced out to rent the movie. "This movie was baaaaad," he wrote, "so bad that even my dog had to leave the room!"

I was particularly interested in several debates on *net.consumers*. Someone asked if anyone else had a particular electric socket in the house where light bulbs seemed to burn out more rapidly than elsewhere. There was an astonishing outpouring of discussion about the economics of long-life bulbs, the relative mechanical stresses on filaments hung upside down, and the light-versus-power efficiencies of various methods of illumination. It typically happens that an expert in the particular subject will join the discussion at an early point and give an authoritative answer brimming with expertise that would appear to end the matter then and there. However, at fist timidly and later in a deluge, mere *people* refute the expert, who is often characterized as having learned this stuff from a book and having no practical experience of the way things really work. Another fascinating discussion ensued when someone naively asked why pilot lights go out. It really had never occurred to me. After all, everyone knows that pilot lights go out, but I hadn't the faintest notion why. (I'm still trying to figure out why rain doesn't come down my chimney.) Anyway, someone from the gas company claimed pilot lights going out was never their fault, and the networking battle was joined.

The most infamous discussion group is *net.flame*, whose motto is "Don't argue over anything that can be settled with a flamethrower." It is said that one of the software architects intended *net.flame* for catharsis; it is used for venting steam over any complaint or perceived injustice. As one user explained the tone of a previous exchange: "The discussion was conducted in *net.flame*, where logic and civility are outlawed." Some entries are isolated cries into the night, such as one person's complaint about actors pretending to play instruments: "Then there are pseudoconductors pseudoconducting a pseudoorchestra, but after all we live in a country where Liberace can give concerts and not be put to death."

On other occasions hundreds of people will chime in to echo a complaint, as when someone wrote how much he disliked the "twits" at a particular fast-food chain who always insisted on asking him if he wouldn't like cheese on his hamburger or didn't he mean to order fries too, or some such order-expanding suggestion. (One solution that was proposed: every time such a question is asked, remove some item of food from the order. Another user offered a fantasy of Clint Eastwood staring silently at the offending salesperson until the original order was filled.) Another popular complaint centered around junk mail that displays the words "You may have already won ..." on the envelope. I'm sure you get the idea.

I personally have learned a lot from the wisdom of the net. In addition to all kinds of valuable computer advice, I've even learned how to protect my car stereo from theft with a small thermonuclear device. (One must add a prominent sticker reading: "If you decide to steal my car or any of its contents, please don't take my expensive and hard-to-replace stereo unit located just to the right of that small nuclear device ..." I am fascinated by the question of whether a new form of social interaction is in the making, or whether the phenomenon is just passing through an interesting, useful phase on its way to oblivion as thousands of new subscribers increase the noise and junk level until nothing but echoes of what-might-have-been remain.

The Gigabit Network: Who Needs It?

A bill before the United States Congress, sponsored by Senator Albert Gore, Jr. (D.-Tenn.), would authorize the construction of a nationwide gigabit network to connect educational and research institutions. The issue that keeps getting raised is: what would a user do with a gigabit data link?

Good question. A few years ago I did not even know what a gigabit was; now I need one of them desperately. Otherwise I may be seriously disadvantaged in this fast-paced information age environment.

Three hundred years ago people got along very nicely with a nationwide *bit* network. Using puffs of smoke, the repeater stations forwarded the digital streams. The technical papers of that day did a thorough analysis of the smoke signal channel. Equations describing the diffusion of the smoke in various atmospheric conditions were studied in depth, and the results were confirmed through extensive propagation experiments. The channel was good for one puff — one bit — per second. Who could ever ask for anything more? One puff could represent war, and two, peace, the code being designed to use the least number of puffs on average. What else could anyone want to know?

Before long, packet communication was invented. Traveling at a small fraction of the speed of light, four-footed common carriers conveyed modest-sized packets of bits between switching nodes. In what was later to become the United States, flow control was exercised by the indigenous natives. The protocol was a simple ACK acknowledgement—a forerunner, literally, of BISYNC. No one complained about latency or error rate. After all, the throughput was as great as the theory of equestrian transmission allowed. Nothing could outrun this optimized system.

At first, business was slow. The usefulness of the new network was not immediately apparent. Why would anyone want to send information westward? However, once the users became accustomed to the network a new category of message was discovered. In addition to "war" and "peace," people found it useful to add the message "send money." The beauties of higher-capacity transmission had been uncovered.

With the invention of the telegraph, speeds were boosted to an unthinkable 10 bits per second. Of course, no single user could make use of such an enormous speed, since it was far beyond the capability of a human to generate messages at this speed. It took considerable time to compose a proper message in the new language of "telegraphese," which used the least number of words to convey a given meaning. Then this carefully composed message had to be hand-carried to the telegraph office, where it instantly disappeared in an barrage of clicks. It was inconceivable that anyone would want a higher transmission rate. What would be the point?

It was only the blink of an eye ago, earthtime, that Marconi sent the first wireless bit. Recently I visited Marconi's home on a hillside near Bologna, Italy. I looked out the second floor window toward the neighboring hill where Marconi's assistant had waited with a primitive receiver for the first click of Marconi's telegraph key. That nearby hill did not seem to me to be very far away, but to Marconi it must have seemed on the other side of the world. When the moment came, Marconi did not say "Come here, Watson." He said "1." Somewhere out there in the universe, that first bit is still traveling, infinitesimally diminished.

I imagined Marconi later staring out his window and thinking that there would be infinite communications capacity for everyone. All that was needed was ether, and it was everywhere, free, and limitless. Yet in the brief years since that historic blink of the eye, we used it up. How could we have been so profligate?

The telephone at first represented a pause in the digital evolution. Not that it mattered, since hardly anyone wanted the telephone anyway — certainly not Western Union, which rejected the telephone patent offered to it for a mere hundred thousand dollars. What user could possibly want this raw capacity in his own hands when expert telegraph operators were there to handle the *real* information medium of text?

Even users were not easily easily convinced of the telephone's value. Early advertisements pleaded with prospective customers:

The proprietors of the Telephone are now prepared to furnish Telephones for the transmission of articulate speech through instruments not more than twenty miles apart. Conversation can easily be carried on after slight practice and with occasional repetition of a word or sentence. On first listening to the Telephone, though the sound is perfectly audible, the articulation seems to be indistinct; but after a few trials the ear becomes accustomed to the peculiar sound.

Eventually, customers discovered that they liked the freedom of expression offered by the new-fangled telephone — so much so that today I am hard-pressed to discover the corporate organization responsible for the second "T" in AT&T. Over the years of this century the function of the telegraph has been subsumed by modems, which initially transmitted at the speed of 300 bits per second. Though later I myself was engaged in the development of higher-speed modems, I had no idea why they would ever be necessary. The speed of 300 bps was exactly the rate that the average college student was able to read. What else was there?

And so it continued. Only now are the 9600-bps modems becoming popular, after perhaps 20 years of availability. It turns out that they are neat for facsimile, and great for file transfer ... Who would have thought?

Now some people shake their heads at the folly of ISDN with its 64-kilobit-per-second channels. There are no applications, they say. Meanwhile, the engineers aren't listening. Wideband ISDN, FDDI, and IEEE 802.6 — all in the neighborhood of 100 megabits per second — are where the action is. A pity that no one will ever need them.

Senator Gore's gigabit network bill is a dramatic strike toward the future. But Congress wants to know: who needs a gigabit, and why? Let the need dictate the evolution, Congress says.

So people are mumbling about access to supercomputers, and talking about the lagging educational system. This conjures up the image of the disadvantaged student in the ghetto doing three-dimensional visualizations of her computational fluid dynamics computations on the Cray, which this wonderful gigabit network has made accessible to all. If this picture does not ring true, it is because we are not very good at predicting uses until the actual service becomes available. I am not too worried; we will think of something when it happens.

Where will this evolution end? Our optic nerves carry several gigabits per second of information to the cerebral cortex for visual processing. How could we ever use more than that? On the other hand, we aren't the only users of a network. There are machines out there too, and who knows what they need? Do you think they would settle for a terabit?

Part 4

Playing the Corporate Game

I was escorting a visitor toward my office, describing the excellence of my company as we walked, when our gazes were arrested by the erratic progress of an approaching engineer. There was, unfortunately, no mistaking his occupation — possibly a scientist, admittedly, but there were no other options. There was no mistaking his *pre*occupation either. As he walked, his eyes stared blankly off into another dimension — surely not one of the three to which we are accustomed. His sole contact with reality was through the back of his left hand, which was dragging against the side of the corridor and guiding him like a seeing-eye dog.

The engineer approached us like a sleepwalker, having no consciousness of the imminent T in the corridor that lay directly in front of him. Suddenly, at the intersection in the corridor his trailing left hand switched him into the cul-de-sac that formed the lefthand junction. He executed the lefthand turn mechanically like a train engine switched into a disused siding, and immediately banged into the blank wall ahead.

My visitor looked at me keenly. "I see that survival of the fittest doesn't apply here," he observed wryly. "Here" is Bell Laboratories — a part of AT&T and formerly a part of the Bell System. Though I have kept that affiliation explicitly out of my column, I have spent my career here, and frankly know no other environment in detail. However, I strongly suspect that the environment for engineers is much the same at all large corporations. Certainly it is an environment full of idiosyncracies engendered by the unique mix of engineering mentality combined with a corporate business culture. Sometimes this is like mixing oil and water, and we should not always expect to have a harmonious relationship between our professional and corporate lives. There is never a lack of material for columns, to be sure!

Management

I was getting out of the elevator one day a few years ago when a stranger turned to me and suggested that I write a column on "the foibles of upper management." The tone of his suggestion might have been described as arch; I got the idea that he had my own foibles in mind. During the decade that I have written "Reflections" I have held a position that many people would describe as *upper management*. Yet in my own view I am hopelessly down the rung of power and responsibility. It is one of those great, never learned, lessons in life that, wherever you happen to be, there is always a *THEY* somewhere mystically above you. It is *THEY* who hold the power and are responsible for all the terrible doings and pronouncements that are a daily part of corporate life. So let us start at the top, or at least as close as we can get.

Who runs these corporations? In the beginning of your career it is almost an irrelevant question. There is somebody "up there" who is all-knowing — somebody so far above yourself that you are like an ant at the foot of Ulysses. The truth is even more disturbing than this; there are actually real people up there. Frightening, isn't it?

I have been able, on occasion, to work with a number of top executives from a variety of corporations, and in almost every instance they have impressed me as being very capable, well-meaning individuals. But the complexities of the business and technological worlds today exceed the grasp of any single human being. Thus we look for giants, but we find only the best of ourselves.

One of the disturbing trends to me as an engineer is that the CEOs of many technologically oriented corporations seem to be coming increasingly from the business side of the house, rather than from the engineering side. This is certainly true in a number of large U.S. corporations, but I have no data to back up this impression. In contrast, the Japanese corporations that make their living from technology appear still to be run by engineers. This impression forms the basis of my column, "The Bean Counters," which is written mostly in the form of a parable about ocean-going boats and their captains.

Whether or not corporations are run by bean counters or engineers, there seems to be a yearning in the engineering ranks for something called *vision.* This yearning to know where we are ultimately headed reminds me of a transatlantic trip I once made on a military transport plane. I was sitting in the rear with the cargo when the young pilot came back to inform me that new calculations showed we lacked sufficient fuel to cross the ocean. We would land instead at Bermuda. This seemed fine with me; my thoughts turned to an afternoon sojourn on a beautiful beach.

A little later the pilot came back to chat again. The headwinds were worse than expected; now we did not have enough fuel to make Bermuda. Not to worry, though, because we could land on an air force base on a small mid-Atlantic island. No problem.

Still later, the pilot came back for a more extended stay. The old general had now taken over control of the airplane, which made me slightly nervous, since the general had had very little sleep the night before. The pilot now confided to me that the crosswinds were too high for us to land at the intended air base. There was silence for a while as the plane droned inexorably westward. Finally I broke the silence. "Where are we going?" I asked. The pilot gave a shrug. "Nobody knows," he said. And so it is in many corporations today. (P.S.: we ended up in Torrejon, near Madrid, Spain. It wasn't New Jersey, as intended, but it was dry land!)

I seem to be writing in parables again. The point is certainly made; people everywhere respond to vision and leadership, but there seems to be less of this commodity available today, whether it is in corporations or governments. Perhaps the world is too complex, and the yearning is simply an unrealistic desire for unobtainable simplicity. Nonetheless, let us return to the question of who runs corporations. If the leaders themselves are not leading, who is? The answer, as everyone knows, is that it is THEM.

Whatever happens at a corporation, it is something that THEY have decided. When rumors sweep through the halls, it is whispered that THEY are thinking of doing such and such. Nobody ever exactly specifies who THEY are. I have a secret answer to this profound question in the column "Corporate Communication," but you will have to read the column in order to gain this deep secret.

As I myself am a corporate manager, this essay reveals my split personality when it comes to corporate communications. Frankly, I am usually the last to know anything; the rumors from the bottom reach me at about the same time as the "truth" from the top. While these bits of information are wending their way toward me from their respective sources, my subordinates are always complaining that they are not being informed. Someone is not telling them whatever it is. Never mind what it is — whatever it is, management is keeping it secret. As a manager I often know that there is nothing to know. Obviously, no one will believe that. There *must* be something to know; management is keeping it secret.

The other half of my personality is that of an underling. (Funny word, *underling;* who would ever want to be an underling?) Anyway, I have this frustration: somebody up there is not telling me what is going on. I know that there is something happening. You can tell that something important is in the works because no one upstairs has been saying anything lately. There is obviously some earth-shattering announcement in the making. It will happen any minute now.

Then there are those rumors that invariably keep life interesting. Rumors surely rate as one of the great mysteries of life to me. Why do they spread so fast and so widely? Why do I myself feel this compulsion to pass on juicy rumors that I feel, intellectually, have no chance of being true? How could we harness the power of the rumor mill to pass on desirable information? Good mysteries all.

Great Myths of Corporate Life

Besides mysteries, corporate life is full of great myths. Some of these myths are unprovable hypotheses that confound rational explanation but which nonetheless are accorded utmost respect. The kind of myth I mean is similar to the old curse of bad luck being caused by walking under a ladder or having a black cat cross your path. Frankly, I never walk under ladders. It is not that I believe in the curse, but why take chances? I notice that just about everyone else behaves in the same way.

One of the most compelling myths is that all demonstrations will break under the scrutiny of high-level management. This is explained in "Management Power." This phenomenon is probably a special case of a more general theorem in life, dealing with the specifics of Murphy's law. Murphy's law, as everyone knows, says that if anything can go wrong, it will. But Murphy failed to pursue his work in various important aspects — a familiar engineering behavior, usually cited by authors of follow-up papers. Anyway, Murphy failed to be specific about the dynamics of his going-wrongness. The question is *when* do things go wrong? The answer is, of course, that things wait to go wrong until the moment when going wrong will cause the most damage.

It is curious that there is a currently popular theory of the origin of life that suggests that life derived from an inorganic material, such as silicon. The argument involves the regular atomic lattice structure of silicon and the possibility of replication of certain structures favored by the Darwinian mechanism of survival of the fittest. Some reputable people have taken to believing in this theory, so it has gained some degree of scientific credibility. However, this theory will come as little surprise to any electrical engineer. We have long known that there is life beating within the heart of silicon.

Designing and constructing a breadboard circuit or a VLSI chip can be like breathing life into an inert form. Suddenly what was nothing becomes sentient. Even as you gain a feeling for this new existence, you sense a certain malevolence. You would think that the circuit would hold some appreciation and respect for its creator, but you quickly come to realize that this will not be the case. Every little circuit has the soul of Frankenstein buried in its psyche; this may become a duel to the death with its creator. Nor should programmers feel immune to this phenomenon; the next biological theory of life may well deal with the replication of data and instruction sets. They will probably give it some acronym like DNA.

This scheming frailty of demonstrations is well known to engineers, but when it is injected into the corporate climate, special problems arise. There seems to be a primitive intelligence in the circuits that senses emanations from management and responds in the most inappropriate way conceivable. Naturally the circuits respond especially to military authority. Eberhardt Rechtin, then president of Aerospace Corporation, wrote me that in the missile/space business he had long used the wisdom that "the probability of launch success is inversely proportional to the weight of the brass in the vicinity." I am sure that many readers remember seeing the German films from WWII that show the generals scattering frantically as the roaring missile settles back on its launch platform.

I have often had the occasion to witness the experimental failure phenomenon from both sides of the bench. In fact, now that I have been a manager for some time, I have come to feel inadequate when I see a demonstration actually succeed. I begin to doubt my managerial demeanor and my aura of power. Sometimes I even see the same demonstration fail for the next visitor, who is a higher-level manager. When even the circuits doubt your authority, where can you turn?

Before I get carried away about demonstrations that fail, I should consider the sorry state of demonstrations that do succeed in these days of hardware computerization and unfathomable software. The old whizbang demos seem to be disappearing in favor of scripted text and graphical output on a computer monitor. Boring stuff, I'm afraid. In contrast, "Dynamic Demos" recounts a colossal failure of a sabotaged demonstration back in the good old days — nothing like the generals scurrying away from the retreating missile, but as close as you can come in a serene electronics lab.

Nowadays it is hard to tell when a software demonstration has actually failed. I look closely at the expression on the demonstrator's face, but people are getting awfully good at concealing these rare failures, which are mostly due to accidental deviations from prepared scripts. Usually I have a hard enough time just trying to figure out some intelligent question to ask. The person showing the software is invariably a virtuoso pianist on the terminal and has spent months playing with the interface, whereas I do not even have time to read the screens as they go through. "See?" they say. "Hmm," I nod ambiguously, thinking that perhaps I can steal back later in the dead of night and play with the system to see what it does. I never do, and I couldn't anyway.

Not long ago I was on a study team looking at the state of some branch of military electronics. As usual, the military organizers had set up a death-defying tour of various indistinguishable laboratories. At the first meeting of the team, however, the chairman of the study endeared himself to me forever when he announced his first rule of task force scheduling. "No demonstrations," he said. Amen, I thought. Sometimes it takes courage to go against conventional wisdom. There are times when, if you've seen one, you've seen them all. Of course, there are other times too.

Technology Strategy

How do things really happen in a corporation? Technology things, that is. Is it good management strategy or just happenstance? Naturally, if you are a manager, you like to think that your farsighted vision has unerringly led research and development in creative and profitable directions. There is, however, another possibility — you may have had nothing to do with it.

In "The Fashionable Thing" I muse about the way we all seem to be caught up in the never ending quest of being fashionable in our technology. Whatever our friends and competitors are doing, that is what we must be doing. Naturally, this narrows the choices that we are allowed to make. It is as if we were all playing a giant game of follow-the-leader. The difference between the game and life, however, is that in life there is no perception of a leader. Picture instead a circular chain of people hanging onto each other, each person blindly following the person in front, who is following the person in front, etc. — the blind leading the blind. Imagine this scene as painted by Hieronymus Bosch and you will visualize what I have in mind.

So who does start the fashions? In the column I duck the issue by depicting the circularity, but sometimes in the dark I wonder. Is it the aliens again? I blame them for a lot of things. "It's time to introduce upside-down lasers," say the aliens. "Just watch the humans scurry about!"

After the publication of this column I got comments from a number of research executives. "Very amusing," they said out loud, but if I could decode the between-the-lines message, it might have been a grudging, "maybe you have something there." Maybe not, though. At a meeting in Washington, D.C., I made the comment from the podium that a certain government report on technology was (merely) an "expression of fashion." Since I was one of the authors, I felt entitled to this interpretation. Unknown to me, however, the very next speaker, a former head of a very large and all-powerful government agency, was the originator of a similar report. He was not amused. He said something to the effect that such reports were the result of a great deal of deliberation and research. And so it is.

At the end of this particular column I said that a *New York Times* article had pronounced that strained quantum-well lasers were fashionable. Unlike upside-down lasers these things do actually exist, and the *Times* article was real. I got a letter from a researcher with a patent on these lasers, wanting assurance that I had stopped kidding at this point in my column. He wanted to be sure that I was taking these lasers seriously. Probably they are not a fashion; they will be forever.

This is a problem I often get into. Many of my columns have a tongue-in-cheek element, but they also have an element of seriousness. Most of them have a real message cloaked in a bit of humor. Sometimes, though, people have a hard time discerning which is which. I get indignant letters from people who don't realize that a whole section is a put-on — like everything has to be deadly serious. Lighten up, I want to say. Interestingly, the computer community has explicitly recognized this dilemma, and many people who post on computer bulletin boards use the special symbol :-) to indicate a "funny face" (view it sideways), meaning that you should take the preceding expression as facetious. Sometimes, though, I favor a bit of ambiguity. It is nice to have latitude in the interpretation of what you meant.

Going back to the game of follow-the-leader, the abject position of hanging onto the person in front of you is not an uplifting strategy for a corporate manager. Many managers have a different game in mind — that of leapfrog. Every time I hear the notion of leapfrogging a competitor I cringe. Even glancing at the article "Leapfrog a Generation" brings on an involuntary cringe. Actually, I do a lot of cringing, because there is always yet another manager who believes that he or she can actually jump over the competition.

The poor competition! Here are a bunch of inept engineers who are going to sit still with old technology while they are over-leaped by the ever-charging good guys from the home team. I don't know how many times I have seen a research result at one company compared against an advertised product from a competitor. "Look how much better we are!" the researchers say. Rationally, everyone knows the comparison is unfair to the extreme, but we always practice a form of self-hypnosis and we end up believing it ourselves. "Yeah for us!" we cheer.

It is not only that the competition obviously has its own research and its own new ideas on the drawing board, but that they also have the advantage of the experience of developing that now-obsolete product. Maybe you have to go *through* the older product in order to reach an informed definition of the next generation. Maybe, now that I think of it, that is why fashion is important after all. Stay with it, or else.

Corporate Information

Why does a corporation invest in research and development? Once upon a time it seemed obvious, but now the whole concept has been called into question. Some years ago I was intrigued by the research of an economist friend who had a theory of the "nonexcludable good." For engineers like you and me, it is easier to talk about lighthouses. A lighthouse is an example of a nonexcludable good. If I decide to build a lighthouse and I get donations from everyone but you, there is no way for me to stop you from using it.

Is research like a lighthouse? Is it something that goes inevitably into the public domain where everyone has free and easy access to it? Or is it something that we can own; something for which we can charge admission? Frankly, I am troubled by the question. Virtually all of the researchers that I know are strongly motivated to publish their work for the world. They want to become famous. They want to have their own lighthouse that every passerby can admire. A little plaque at the bottom of the lighthouse would have their name enshrined forever. No, that is not correct — it would have to be a *large* plaque. "Oh, look at Dick's lighthouse!" people would say.

Unfortunately, corporations must make a profit. Constructing free lighthouses does not offer great appeal to business managers. Thus in recent years there has been a growth in a new profession devoted to intellectual property. These people specialize in the construction methodologies for private lighthouses, edifices that emit no visible glow in public directions. The only way that their lighthouses can be admired is through the payment of an admission fee. Whether or not such lighthouses can be built is the subject of "Proprietary Information."

I have to confess that the flow of information in the technological world is one of the many things that I do not understand. I often think that someone should study this so we could all appreciate from whence information comes and to whence it goes. However, I also realize that, not only is the subject too complex for comprehension, but the world changes faster than any possible rate at which our understanding could grow. In other words, I consider it unknowable.

The lay public has this concept of technological innovation where an engineer is soaking in a bathtub and suddenly cries, "Eureka! I've discovered the flogistor!" At a flash a whole new field is created from whole cloth. I certainly have seen cases where this is approximately true, but it seems to me that virtually every invention has a tangled history of predecessors without which the present innovation would have been impossible. In other words, you will not be able to build your private lighthouse without the illumination of many other lighthouses. Catch-22.

History tends to simplify events. Did you know that the invention of the telephone by Alexander Graham Bell was a 5-4 decision of the Supreme Court? If one justice had voted the other way, someone else would have been given the patent on the telephone. Do you know who that someone was? As they say, close only counts in hand grenades.

All of the good researchers I know are bit players in an international drama. They peck away at the edges of the information frontier, always conscious of the ever-moving boundary of what is known and what isn't. Working alongside them are researchers from England, Germany, Japan — wherever. The idea that each of these workers would wear intellectual property blinders so as not to be aware of what co-workers were doing seems ludicrous. The world's research system only works effectively by an open sharing of information.

This argument assumes that the aim of research is an altruistic one of helping mankind. The problem is that every worker in that international field must have his or her salary paid by an entity with some private or national interest. How do we reconcile these interests? Does the answer lie in new and better laws for the ownership of intellectual property? Frankly, I do not know, and any bias the reader may detect is not strong. I would prefer a less wimpy position.

Supporting Professional Societies

One of the rewarding aspects of engineering is that we have an affiliation larger than our companies — the engineering professional societies. These societies glue us together in a larger sense, both technically and socially. They give us the forum and the mechanisms to share that information whose generation has been funded by our employers. Usually the relationship between employers and societies is symbiotic, but there are constant little questions about the degree of support to professional societies that employers ought to give.

In my experience two of the most recurring and heartfelt questions have to do with the clearance of publications and the attendance at certain conferences and symposia. People get really uptight about these things, and I often wonder whether the few restrictions that industry places on them are even worth all the fuss. I have already mentioned how important it is to researchers to have their papers published. They want to become famous and get lots of high-paying job offers so they can leave their employer, who never appreciates them as much as they should be appreciated. No need to go into that again, so let us turn to the other generic problem, attendance at technical meetings.

"Tahiti or O'Hare?" crystallizes the dilemma. Contrary to popular belief, engineers like to have fun. But engineers usually have a streak of puritanism and a strong work ethic, and these two aims periodically come into conflict in the choice of venue and style for our professional conferences. I like to go to nice places as much as anyone else. In fact, I am writing these very words from a hotel room balcony overlooking swaying palm trees and blue ocean. I am waiting for the technical sessions to begin, and I feel guilty.

Just yesterday I received a request for foreign travel from one of my engineers to go to the French Riviera, exactly similar to the request that triggered my column about Tahiti and O'Hare. I hesitated long over signing the request, and only the sure knowledge that I would alienate the engineer irrevocably for a paltry amount of money forced me to sign the permission form. I am not sure whether I felt jealous, or whether I felt that the company would be wasting its money in a frivolous manner.

Worse than just wasting the company's money would be the public display of unwise largesse. I visualize the headline, "Engineers frolic on Riviera beach while layoffs in factory persist." Yet it is entirely possible that the conference in the Riviera is vitally important to the education and networking of that engineer. This is a dilemma that occurs time and again, and causes all of us to be careful in the selection of sites for our meetings. Careful up to a point, that is. I mean, nobody wants to go to a meeting at O'Hare Airport in the dead of winter.

We usually work at meetings, and we network, and we have some fun. Actually, I would like to think that we work harder than happens at the conferences in other occupations. Not long ago there was a television report on a doctor's convention where they interviewed doctors on the ski slopes while the meetings were going on. Bad press. I want to tell a little story that reflects well on engineering conventions, but I hope that not too many readers will take offense. Here goes.

This was a large IEEE convention in a large city. It was held in one of those hotels with interior glass elevators that rise high above the chasm that forms the central lobby. It was moderately late at night, and I was returning to my room after a dinner out in the town with some of my engineer friends. I got into the glass elevator by myself, but just as the doors began to close, in walked two extraordinarily attractive young women. They were dressed in revealing outfits and drenched in perfume. What can I say? It was obvious even to a naive engineer that they were working in their own profession. There was silence in the elevator for awhile as I mentally whistled to prove my indifference. Finally one woman turned to the other. "How are you doing tonight?" she asked. The other one shook her head in disgust. "This is the last IEEE convention that I'm ever going to work," she said. I haven't seen them or any of their friends since.

People sometimes ask how they can get involved with IEEE work, and I always recommend joining some committee or other. There are a lot of committees that are just looking for people willing to do a little work, and the rewards for becoming a part of the fabric of the society are deep and lifelong. It is certainly easy to poke fun at committees in general, and I do a little of that in "Committees and Canines," but I believe that the social value of the collaborative work in a committee outweighs its inefficiency as an organizational entity.

For many engineers their actual work is by its nature isolating. The depth and complexity of our problems exclude others. Often it becomes too laborious and time-consuming just to explain what we are doing to a fellow specialist. Working in a committee, whether it is a corporate committee or a professional committee, gives us a chance to work shoulder-to-shoulder with other engineers on problems that are usually more human than technical. There is something about working together with shared objectives that creates lasting friendship, and I value all the friends throughout the world that I have met and learned to like through professional committees.

I remember well the first week after I received a significant promotion to a nearby, but different, organization within my company. The thing I most remember is that I had no one with whom to eat lunch; my old lunch gang was no longer accessible. In all the years since, I have never had a regular lunch group. No manager that I know does. Being a manager, too, is isolating. Some of this feeling helps explain why even CEOs of large corporations often like to join working committees. In my years as chairman of the Air Force Scientific Advisory Board I came to understand how famous and important people loved to join study teams and to work on problems that were often grubby. I suspect that the truth was that they seldom got the chance in their normal jobs of doing any meaningful work at all, and never had the opportunity to work with peers on technical stuff. They enjoyed it, as would any of us. The only problem is that you do not get a million dollar salary for doing this kind of thing full time.

Engineering Pride and Pigheadedness

"Just like an engineer" is an epithet that I hear a lot. There are certain characteristics of engineers that are well known to nonengineers. We know them all too well ourselves. In engineering school we were all trained in the formative years of our lives to have certain values, not all of which serve us well in our subsequent careers. Someone once commented to me that engineering is the only profession that is taught by nonpractitioners. Think about that.

Anyway, at the top of the list of values is individual creativity. The engineer at the top of the heap is the one who invents the most. You can get to the top of the heap in two ways: by inventing more yourself, or by having others invent less. The first is probably harder than the second, although it is hard work continually finding fault with everyone else's ideas. Nevertheless, we must persevere in this difficult compulsion. "Too bad nobody around here has any good ideas except me," we tell ourselves. We even believe it.

Two of my columns, "Turf" and "That Dreaded NIH Syndrome," deal with this tendency of engineers to put down the ideas of their fellows. In my company the engineers are famous for this. I find it in myself all the time. Someone will come to me with a new idea. I hear them out, all the while calculating what is wrong with what they are saying. I mean, it couldn't be any good, could it? After they finish, I smile condescendingly. "Have you considered what happens when the freeble is backwards?" I ask. Usually they have, as it happens, but that is just my first line of defense. I don't usually expect such a simple ploy to work. Before long I am resorting to what I always think of as the "Pliny the Elder criticism." Even if their idea is good, someone has already published it. Pliny the Elder always comes to mind. Everyone knows that he did a lot, but no one knows what it was.

NIH (not invented here) as practiced in contemporary engineering can be played as an individual competition or on various organizational levels. The column about turf is about "us" and "they." I had just read Robert Ardrey's *The Territorial Imperative* and kept thinking how much like apes we engineers were. Also at that time the movie *Planet of the Apes* was popular, and I could not resist placing my engineer friends inside those ape masks. Nobody could tell the difference through our behavior. "That's an ape," someone would say. "No, it's an engineer," would be the retort. We all have our territory staked out, first for ourselves, then for our group, then for the next higher organization to which we belong, etc. *Turf wars,* as they are known, are an everyday part of corporate life.

A closely related engineering strategem that has been perfected by many corporate types is that of strategic self-protection, known colloquially as CYA. This is discussed in the essay "Finger Pointing," which contains a deeply intellectual discussion of risk assessment, risk avoidance, and risk assignment.

As you undoubtedly know, risk is at the heart of corporate life. Upper level managers will tell you that the name of the game is taking risks. Often they beat their chests when they make these statements. (Whoops, that was the ape column!) Taking risks is macho, it's where it's at. Stuff like that. Anyway, at the time it sounds good. Everyone runs around with bright eyes, breathing fire, and talking corporatese.

This state of elevated animation is, of course, in the beginning of a project. The comprehension slowly grows that taking risks is not necessarily good in itself. *Succeeding* is good in itself. Did you ever hear anyone praised extraordinarily for taking a risk and failing? "There goes Wally," someone says. "You really have to admire him — he took a fantastic risk. Failed miserably, of course. But look at the way he limps down the aisle as if nothing happened. Really hangs in there well, in spite of how everyone avoids him. Yes, you really have to admire old Wally for taking that risk."

Everyone knows that there is a time in the history of any project for the identification of the guilty and the assignment of punishment. No one will receive an extra large raise for having accepted the risk of joining the doomed project in the first place. The obvious strategy is not to be there when the project goes down the tubes. If you are, the fallback position is that it was not your fault. That is where the art of finger pointing comes to the fore. In a modern corporation there are so many organizations with partial responsibility that the reassignment of blame can be quite effective. "They did it," say the two groups simultaneously, with fingers pointed at each other like dueling revolvers. Just one happy family.

The Bean Counters

Once upon a time great ships of state sailed the seas of commerce. On the bridges of these great ships were the captains of industry, the wise and visionary leaders of that day. Many of them were engineers, who personally embodied a breadth of knowledge and wisdom that enabled them to navigate the murky waters unerringly toward the distant lands upon which their eyes were so firmly set.

Those were the days of yore — a legendary season in history when engineers were in charge. But now the mists of remembrance have all but closed over those times gone by. Today the stormy oceans of commerce are full of small and large boats sailing aimlessly under many different flags.

If you have visited the bridge of your boat lately, whom did you encounter there? Did the captain and mates have their eyes turned toward the still-unseen distant lands ahead? Or was their gaze instead cast downward upon the foaming whitecaps just beyond the bow? Or perhaps they were fixated not upon the sea ahead, but backward toward the boat's built-in wall safe. Were the captain and mates, perchance, wearing telltale, three-piece suits? If so, then you have met the new type of business leader — people who are guided only by the short-term accumulation of money. Since talking about money is crass, we call them *bean counters.*

In the days of old the crew of the ship would gather around the captain to listen to his tales of the great land beyond — the joys and benefits that would ultimately compensate for the hardships of the long voyage. The echoes of those wonderous tales still resonate about us, but those who tell them no longer abound. Now, should a passenger or crew member ask the captain about the final destination of the voyage, the same answer reverberates from boat to boat. "Wherever there are beans," intones the chorus of new captains. "But where shall we steer?" asks the crew, pleading for a motivating vision. "Toward the beans," comes the predictable wisdom from above.

"What happened to the wonderful land beyond that our old captain promised?" asks the crew. The captain frowns. "There were no beans there," he says in disgust. "He was court-martialed. We don't allow people like that to command boats anymore. Now get back to work."

The engineers have now been relegated to the engine room below deck, where there is no sunlight or any notion about where the great ship is headed. Still, they take great pride in the gleaming new engines, which they continually improve and polish. "Does the captain know the potential capabilities of our wonderful new engines?" asks one of the faceless engineers. An equally faceless companion looks up from her engine-software. "How could he?" says the software designer with resignation. "He has never been down here."

Dispirited, but determined, the engineers discuss among themselves the possible new lands that might lie out in the ocean, if only there were resources for certain improvements in the drive and navigation equipment. They decide to commission a small group to approach the august captain to plead for their plan. Unfortunately, the captain is too busy to meet with the engineers. He has a full calendar of appointments with various people who specialize in predicting bean crops. The humble engineers are not surprised; they had only expected as much. None among them had ever met the captain, though they had often seen his picture in the popular magazine *Beaness Week*.

An assistant to the first mate is delegated to meet with the engineers. He frowns his displeasure at their appearance — conspicuously sloppy in comparison with his own. Though he himself is only an apprentice bean counter, he keeps his shoes well-polished and his vest fully buttoned. Surely it cannot be long before he himself takes command of a ship. After all, he has a master's degree in bean administration from a great school with alumni in important positions throughout the bean economy. As for the dress of the engineers — what can you expect? They have spent too long in the engine room; they know nothing of the real world.

The engineers explain to the assistant their plans for an upgrade of the equipment of the engine room. The assistant is mildly amused at the computer-prepared charts that the engineers show. Typical, he thinks, even to the inclusion of several mysterious-looking equations that the engineers seem to think justify striking out in a new direction. His face set in an expression of impatience and concern, he daydreams until the engineers reach their final chart — a spreadsheet showing the accumulation of a future crop of beans.

Now he pounces on the hapless engineers. "You know nothing of beans!" he shouts. "The last time you made a proposal like this we had a crop failure! It was only through the efforts of our great new captain that we were able to obtain new bean seed. Even if your plans made sense (his expression one of incredulity), you should know that there are insufficient beans in this year's allocation for your proposal. Furthermore, the harvest this quarter is down. This is not a time to bother the captain about such matters."

The engineers hang their heads in disappointment, remembering also the land that was sold, and the farm workers who were laid off. "Why does the world have to revolve around beans?" they think despairingly.

So ends the parable, but how has this turn of events come about? How could mere bean counters, ignorant of technology, take over complex, technologically oriented businesses? Perhaps there is a clue in Jerzy Kosinski's book *Being There,* in which an uneducated gardener, Chance, is struck by a car and suffers amnesia. Treated as an unknown savant by his benefactors, Chance meets the President of the United States. Chance is taciturn, but the President urges him to state his opinion of the current situation in the financial markets. Chance replies in the only terms he knows: "In a garden growth has its seasons," he says. "There are spring and summer, but there are also fall and winter. And then spring and summer again. As long as the roots are not severed all is well and all will be well."

Great wisdom is read into these words, and Chance, now Chauncey Gardiner, becomes a government advisor and a media celebrity, as he endlessly repeats homilies about the raising of crops.

Is it possible that we engineers have ceded our futures to a rising legion of Chauncey Gardiners? Have we accorded too much wisdom and too much power to those who have the skills for tending a garden of beans? Or is counting beans a lot more important than engineering vision in the world today?

Corporate Communication

Ever since I can remember, people have been complaining about the lack of communication within whatever organization I was associated with. I myself add periodically to the din, oscillating between "They never tell me anything" and "They never listen," depending upon which side of the fence I am on. Many of us claim to know a lot about electrical communications, but how well do we understand how information is propagated in an organization filled with real people?

Somehow we all have the comforting illusion that there is someone up there who knows what is going on. Usually this someone is known by the dreaded, generic name THEY. Our gripe is that THEY don't keep us sufficiently well informed. We expect that there should be frequent announcements from a loudspeaker outside our office, proclaiming "Now hear this ..." in a stentorian electronic voice. The voice would go on to proclaim the latest bit of strategic thinking from the top. "We are now at war with Oceania," it would continue. "We have always been at war with Oceania." How comforting it would be to be kept informed!

There are all kinds of things wrong with this little fantasy. First, I know that most readers will not truly believe this, but there is no THEY. I hate to destroy Santa Claus, but I have even spoken to THEM, and they say that they are not THEY. Worse yet, they complain peevishly that there is some other THEY far beyond their own pathetic reach. This has ominous overtones. Clearly, the whole concept of THEM is an infinite hierarchy, with the upper levels ascending into a metaphysical fog, and perhaps even wrapping around circularly.

Not only is there not a THEY, but often there is not even an IT to be proclaimed. We all imagine that there is some important news that needs communicating, but what if there isn't? Imagine the loudspeaker announcing authoritatively: "Now hear this. We are currently not sure whether we are at war with Oceania or Eurasia." This is closer to the usual situation, but how does it make us feel? It reminds me of the 19th-century German chancellor Otto von Bismarck's admonition about law-making: "Law is like a sausage; if you like it, you don't watch it being made."

Admittedly, there are occasions when there is a decision to be communicated to the masses below. Fortunately, there is no loudspeaker. Instead, a mute desk-to-desk notice appears with the latest momentous proclamation. "There has been a change in regulation 3101-5b, further defining the concept of a dependent with respect to medical benefits in the case of work-related occurrences of gray plague." Later, should this occasion actually arise, the victim will claim that he was never told. Some people never get the word. This is a general theorem that requires further study.

When something important, like a reorganization, happens where I work, it appears on the bulletin boards. Early in the morning a clerical employee unlocks the sliding glass door of a cabinet and pins up an innocuous-looking notice that may profoundly change people's lives. I used to have the idea that this employee was running the company. I had the fantasy of getting a copy of the key and typing up my own notices, gradually reorganizing the company so as to increase my own responsibility. I can imagine people passively accepting the changes, and with a shrug of their shoulders undertaking their new assignments. How long do you think I could get away with this?

Another way to disseminate information within an organization is through the organization's own newspaper. Most issues of these papers have a familiar look. There is the obligatory picture of the president and maybe a vice president or two. Important people in the organization, speaking through ghostwriters, say how well things are going. Nobody believes any of it — even the true parts. There is a strange discrepancy between these glad tidings and the business news in the national newspapers.

Here I have this fantasy of starting my own company newspaper. It would be sold in stands outside the doors at quitting time. My scandal sheet would feature the lurid rumors about what was *really* behind the latest reorganizations. People would line up to buy the paper, and would believe implicitly all its fictions. Such is life — it is so hard to communicate factual information, but rumors spread like brushfire.

I have always been fascinated with the disparity in speed between rumors and proclamations. The exponential broadcast of rumors is an example of human networking at its most effective. Someone should study this phenomenon: there must be more propagation here than could be predicted by Maxwell's equations. Recently, for example, I observed two rumors sweep nationwide throughout my company within a day or so, even reaching out to remote homes of retired employees. Neither had a shred of validity; maybe that is why they spread so fast.

I myself stay delicately tuned, ear to the ground listening for the distant, faint sounds of an approaching rumor. Curiously, I must confess even to passing on some of those rumors that I believe to be ill-founded. There is an irresistible compulsion to spread rumors; the urge exceeds even that of the threat of dire consequences contained in a chain letter. I throw away chain letters, but I worry. And I think ill of the anonymous people who have foisted this impending bad luck upon me.

Now as everyone knows, the most credible rumors are those arising from the secretarial pool. Herein lies the key to the corporate communications riddle; perhaps it is even a clue to the origin of corporate life itself. In the beginning, there was a secretarial pool. So maybe this is where THEY are. Every night, the secretaries huddle to decide what rumors they will start the next day. A reorganization might be in order, so an appropriate rumor is crafted. It has to be just right, if it is to become reality. Within a day or two, the impending reorganization is being discussed throughout the company. Before long, to no one's surprise, it is confirmed. One governing member of the secret secretarial council winks covertly to another. What a great feeling it must be to be one of THEM!

Management Power

I suppose every profession has its inner secrets, known to all its members but never even whispered about in public. In electrical engineering one such dark truth is the power of upper management to influence the behavior of inanimate objects, such as prototype circuits and systems — always, of course, for the worse.

This basic truth is not taught in school, so it often comes as a devastating shock to the neophyte engineer to see the proudest creation fail at its first exposure to the near field of a Member of Upper Management (MUM). It is understood that within an organization, managers have dominion over man and beast, but how did circuits come to share such a subservient role? Give them so much as a whiff of a MUM and they roll over belly-up and give the most convincing performance that they are not just recently deceased, but have never worked at all.

I do not know when this effect was first observed, but it certainly goes back at least thirty-five years, as it is part of a small legend that has long fascinated me. In the early 1950s D.W. Hagelbarger built a mind-reading machine at Bell Labs. Now I can hear you thinking: "Hoax." But not so. In some 9795 total matches against humans in a penny-flipping game, the machine was able to correctly predict the person's call of "heads" or "tails" 5218 times — a performance whose probability by chance alone would be less than one in ten billion. Much as I hesitate to demystify this technological feat, the fact is that the machine took advantage of the difficulty that people have in making up random numbers, and their sequences of "random calls" have an innate predictability.

News of the machine's omniscience spread quickly, and researchers lined up to try their hand at defeating this mechanical threat. Eventually an order was issued that this champion machine be brought before the director of research, who in case there is any doubt, was a fully qualified MUM. Today's reader will not be the least surprised to hear that the machine was soundly thrashed in competition with the MUM. The principle has long since been established.

But in that yesteryear, while the MUM basked in his confirmed superiority, Hagelbarger, in the privacy of his own lab, replayed the MUM's recorded sequence of calls back into the machine. The machine won every time; it had simply thrown the previous games. So it goes. (Ultimately, Hagelbarger's machine suffered a close defeat in a duel to the death with a similar one constructed by Claude Shannon.)

Although we all take for granted that experimental systems can be intimidated by upper management, it is not known how the effect is initiated. Is there a power field about the MUM, or does the system sense the fear in its builder? Clearly some definitive research is needed. For example, does shielding the system in a Faraday cage suppress the effect? Are obvious precautions — such as keeping one's back to the circuit and having bystanders push any necessary switches — effective in preventing any contagion of fright?

Even though the causes of the MUM effect are poorly understood, there are nevertheless applications that might be exploited. One such immediate application would be the life-testing and burn-in of components. I can imagine running a small conveyor belt through the office of the CEO. Hour after hour the latest components or subsystems would snake silently by the CEO's desk for maximum exposure to the emanations. Naturally, there would be great fall-out during the test, but those circuits that passed would be guaranteed to be of the highest hardness.

What the circuits would endure in their short journey would be similar to what I believe my car endures every time I drive through the Lincoln Tunnel to New York City. (Other readers may substitute local hazards where turning back or off is impossible or ill-advised, like the Golden Gate Bridge.) The principle is that this is the period of greatest psychological stress on the car because the consequences of failure are maximum; therefore the failure rate itself is maximized.

Turning the stress effect to our advantage, we could use its relative strength to select potential MUMs. This might be more accurate than the conventional means. For example, to choose the next IEEE president we could build the simplest and most reliable circuit we know — perhaps a ring oscillator — and then expose it to every would-be president to see who could make the circuit stop functioning. The successful contender would be hailed as King of the IEEE. After all, pulling swords out of stones has gone out of style. We need a modern replacement.

Unfortunately, the "inanimate objects" that are intimidated by upper management often include people. I remember with embarrassment many of my own lesser performances. (Very lesser, actually.) Once early in my career I was called before a very upper MUM on a Friday afternoon. I had never been in his office before. He told me that the next Monday morning there was to be a visit by the CHAIRMAN OF THE BOARD. It was short notice, but the MUM wanted me to give a talk.

Needless to say, I spent a pleasurable weekend contemplating my good fortune. Monday morning I was up early. (I am sure you understand what I mean by "early.") When I arrived at the office two MUMs stared at me nervously. I had not previously realized that MUMs could be nervous. "You can't let him give the talk," the alien MUM said to my local MUM. They proceeded to argue over whether I was to be allowed to give this talk — as if I were not even present. Naturally this filled me with great confidence.

In the middle of the argument someone whom I recall only as a court jester came running into the room. "The Chief's not coming!" he gasped. I wilted with the reprieve, but both MUMs turned threateningly to me. "Let's hear it anyway," they demanded. Well, I have since given many talks under many circumstances, but this was my worst. One might say that I lowered myself to the occasion. In the silence that followed my pathetic presentation the MUMs turned to each other. "It's a good thing the Chief didn't come," was all they said. I have never since given a talk before the chairman. As I reflect on it, not even the actual presence of such an august personage is required to set off panic. The mere intimation of an oncoming appearance is enough to suppress mortals. No wonder the poor circuits cower.

Curiously, many MUMs do not acknowledge their power. "Shucks, it's only me," they say disarmingly to themselves, not believing it for a moment. But legend is full of stories of buildings moved because of a passing offhand remark by a MUM. "Quite a walk from my parking space," muttered under the breath, and miracles ensue. In turn, non-MUMs tend to read too much wisdom into MUM pronouncements. "What do you suppose she meant by 'Good job'?" they worry. "She didn't say 'great job' or 'very good job.' Do you think she is trying to tell me my work is mediocre?"

You see, there is another dark secret; MUMs do not always know what they are talking about. Don't tell anyone. Maybe the circuits don't realize this either.

Toward Dynamic Demos

Presumably young Alexander Graham Bell awoke feeling nervous on June 23, 1876. On that day he was to make history by running the first demo ever where the audience was instructed to listen at an earpiece. Possibly General George Custer also awoke feeling uneasy, for he too was to conduct a demonstration of one of his theories on that very same day.

At the Centennial Exposition in Philadelphia, Bell spoke one of Hamlet's soliloquies into the primitive microphone. Emperor Don Pedro II of Brazil, one of a panel of learned judges, heard Shakespeare's words "To be, or not to be" emanate disembodied from his earphone. "My God, it speaks!" he cried out in Portuguese. Meanwhile, back at the Little Bighorn River in Montana, General Custer's demonstration of his theory of crowd control was not working so well. He was discovering that it was going to be a very bad day.

It is more than a century later, but I have lost count of the number of times that I have been privileged or coerced into playing the role of Emperor Pedro. When I was very little, my brother ran a taut string between two cylindrical, cardboard oatmeal boxes. "Listen here," he said with pride. Later, when I was a teenager, my next-door neighbor got a pair of dime-store walkie-talkies for Christmas. "Just put your ear here and wait while I run across the street," he pleaded.

These experiences are of the everyday sort, but here I am a professional in the telephone business. Time after time someone has pointed to a rack of experimental equipment connected to a set of telephones and asked me to pick one up and listen. Whatever electronics are in the rack are going to do something risky and wonderful, or demanding and dreadful, to the precious voice signal. Perhaps the voice is going to be coded in some diabolically clever way, transmitted through some novel and uncooperative medium, or switched in some death-defying manner. The obvious pride of the designers tells me more than anything else that whatever is being accomplished is indeed a good trick.

Dutifully, I put the receiver to my ear. "Do you hear me now?" they always ask. I always do hear them, and it always sounds remarkably like a telephone call. "My God, it speaks!" I cry in Portuguese. But only in my mind — the truth is that I haven't the foggiest idea how I am supposed to react. Only a faithful dog can display such renascent enthusiasm over this repetitious event. Night after night the dog leaps for joy at the magic return of its owner through the front door. "Fantastic! He's done it again; he's materialized out of the door! What a trick!" thinks the dog. I haven't the dog's guileless perserverence. There must be another demo in this business. The problem is, young Alex Bell didn't just invent the telephone — he used up the only good demo in the whole field.

Back in Bell's day demos were demos, not the wimpy, intellectual charades we artlessly practice today. Imagine Edison throwing the switch on his lightbulb, or Franklin calling upon the heavens themselves with his kite. But what have we today? Recently I witnessed a demo of a complex array processor. The processor consisted of boards filled with microchips that looked exactly like all the other boards filled with microchips from all the other demos. The designer of this system sucked in a deep breath and typed, "Go." After a scant heartbeat, the display responded "Done." Like Edison's bulb, the designer glowed at me. "Wasn't that wonderful?" his eyes beamed. "Done what?" I wondered. And how do I really know that it did what it was supposed to have done? And I suppose that I have to take on faith that whatever was done was actually done by that anonymous stack of silicon slabs.

Proper appreciation of a demo depends upon intimate knowledge of the function to be performed and complete faith in the integrity of the demonstrator. The first of these prerequisites almost never holds true, and even the second is not without question. Every engineer knows that demos are occasionally faked, if only in subtle ways that invoke only the weakest feelings of guilt. For example, there may be that ubiquitous wire under the table that provides timing to the receiver, or various critical parameters may have been hand-tweaked seconds earlier, or the input script for the software may have been especially well chosen. I always wonder if the demo really does what is claimed, and, if so, if it does anything else.

Years ago I was scheduled to give a demo to a group of visiting don't-cares. The prototype equipment I wanted to demonstrate was being revised in the lab, and I was reluctant to interrupt the hard-working technicians. I had a brilliant idea. I would haul a *different*, arbitrarily chosen rack of equipment into the briefing room. Since all electronics looks the same, no one would know that it was not what was claimed. I would then hook this rack of junk to an oscilloscope and throw the power switch. Nothing would happen, naturally, and I would simply apologize for the unexpected failure of the demo. Form would have been satisfied, and everyone would be reasonably happy.

Unknown to me, however, the "arbitrary" equipment supplied to me had somehow been sabotaged. When I threw the power switch I was shocked to hear an ominous, descending siren emanate from the rack, followed by an inevitable explosion. The room filled with smoke, and an obscene, mechanical, cackling laughter was emitted from the charred remains of the demo. To the stunned audience, my apologetic "It didn't work" was superfluous. General Custer would have been proud.

It is not just a question of better show business. Demos are vital to the selling of our work and of ourselves. As an engineer I too have been frustrated by the ever-deepening complexity of our labor, and by the increasing difficulty in bringing to the surface demonstrable effects worthy or representative of the monumental labor submerged within the design process.

This is particularly the case in software, where scores of person-years of creative, intellectual labor are belied by the production of an apparently trivial external effect. How can we show how hard we have worked? How can we share the exultation of success when something that is so complex that it has no right to work at all actually does work? "Listen here," just isn't enough anymore. We're better than that.

The Fashionable Thing

I sat in the darkened auditorium, listening to a talk about the history of fiber optics. It was being given from the perspective of the local telephone companies — British, in this case. The projector was showing these beautiful, old, black-and-white photographs of systems experiments using a circular waveguide, the technology being pursued in the early 1960s for high-capacity transmission.

For those who may never have heard of this technique, it used a hollow pipe of about 10-cm diameter to transmit microwave signals in a mode that had extremely low attenuation — that is, as long as the pipe was absolutely straight. Well, nothing was perfect, after all.

The old slides shifted to another technology — the confocal lens. Now the pipe carried lightwave signals, which were focused periodically by arrays of lenses. The pipe still had to be ever so straight, and there were these other problems ....

I remembered that stuff, and the old-fashioned texture of the ancient slides made me feel like a fossil. What dumb ideas they were. How could we ever have worked on such stupid technology? It was bad enough that we struggled on that junk on our side of the Atlantic, but in Europe they were doing the same thing. So was everybody else, I guess. It was the fashion then.

It is curious how technology travels. As I go from lab to lab around the world, I see the same projects everywhere. A story starts running that

NEC in Japan is working on upside-down lasers. Whispered conferences are held in Yorktown Heights, Murray Hill, Martlesham Heath, and other famous electro-metropolises.

"They say it has a fantastically high output power," says an electro-manager in a tense voice. "We have nothing like it," he adds, needlessly. The assembled researchers stare quietly at their notebooks, edging nearer to each other in order to share the collective guilt. One notebook seems to have a diagram of an upside-down laser drawn in freehand. Or perhaps the notebook itself is upside-down. In any event, later that afternoon eleven people are working on upside-down lasers.

Several weeks later at the National Science Foundation in Washington, D.C., a staff meeting is taking place. "You mean to say that we have no program in upside-down lasers?" the director is saying quietly through clenched teeth. Now the staff cringes with his sudden rise in volume. The director is shouting the ultimate threat. "What will happen to national competitiveness!?" he exclaims.

Special funds are quickly allocated at NSF. The word goes out: there is money on the street. Just say the magic words — upside-down lasers. In Cambridge, Pasadena, Palo Alto, and other electro-villages the sweet smell of available funds permeates the air. Ph.D. students confer in hallways. Upside-down lasers are fashionable. Get in while you can.

Meanwhile, in Tsukuba, Japan, there is a meeting of executives at the NEC labs. "Do we have anyone working on these — what do they call them? — upside-down lasers?" asks the chief manager. The others stir uneasily in their seats. "I regret, but, no," says the group leader. "It is an American thing," he adds with a helpless shrug. He glances sideways at the others with an expression that seems to say, "What can you do?"

Nevertheless, a team is quickly assembled, and a leapfrog project is begun. If the Americans want to push upside-down lasers, they will find the Japanese more than ready.

At the next Conference on Lasers and Electro-optics, the sessions are filled with upside-down laser papers. Reporters for trade magazines buttonhole attendees outside the meeting rooms to get interviews for feature stories on upside-down lasers, but even those agile publications are scooped by an in-depth story in the *New York Times* science section.

In the midst of all this clamor, a quiet rumor is being circulated in the corridors of yet another breakthrough in laser structures — this time, the sideways laser from Stuttgart. No one knows the details, but it is said that this laser has an incredibly low threshold current.

Later, back in New Jersey, the vice president looks over the long table in the executive conference room at the upturned faces of the directors. "Why are so many people working on upside-down lasers?" he asks in exasperation.

A director not responsible for laser work chimes in with the news about the German success with sideways lasers. The other directors are irked by his smug expression; they knew, too, and why bring up this old news at the staff meeting?

The director who is responsible for laser research seems to be drawing abstract doodles on his notepad. Perhaps it is a sideways laser he is drawing, but it could be that the notepad itself is sideways. Whichever, it is people as well as lasers that are about to be reoriented; fashion has changed. The story in the *New York Times*, quoting everyone about the lasting greatness of upside-down lasers, is now an embarrassment.

The history of technology is a tangled one, full of backtracking and wrong turns. Surely, that is to be expected, but the surprising thing to me is that we all seem to follow that tortuous path together in simultaneous misadventure.

Everyone and every organization is afraid of being left out, like a child not belonging to the right clique. There is a fashion in technology, like the fashions in styles and clothes. That irresistible fashion travels across the world as an undercurrent with the speed of light.

I remember being worried years ago that my company did not have a strong position in fluidic logic or in magnetorestrictive delay lines. Not so long ago, we were scooped on cold fusion, and were running hard to keep up with high-temperature superconductivity.

Fortunately, all that is in the past. Today we are on a firm course. The *New York Times* tells me that strained quantum-well lasers are now what matters.

Leapfrog a Generation!

On several occasions I have watched a college football game between, say, Southeastern Middle Nowhere and Techless Tech. Techless Tech, in spite of a thus far winless season, holds a slight lead in the third quarter. As the TV cameras focus on their bench the players jubilantly hold their index fingers high, signifying, "We're number one!" "Who do these turkeys think they are?" I muse to myself. Yet the same thing happens in electronics. Everybody has to be number one. It is a good thing there are so many openings in the world for number ones; otherwise it would be very crowded at the pinnacle.

There is a meeting at Company A. The purpose of the meeting is to discuss strategy in the competition with Company B. Lest there be any doubt, Company A consists of the good people, whereas Company B has obtained its competitive position through underhanded dealings and blind luck. Be that as it may, the managers at Company A admit to themselves that the product in question from Company B has a better price:performance ratio than their own. This admission is a magnanimous one, in the best spirit of their business school training—besides which it is well known that a product gap is never attributable to upper management. Obviously it is not the fault of the sales force, which is doing exceptionally well with such an inferior product, nor of the marketing people, whose advice had been thoughtlessly rejected. No, it is obviously the fault of the incompetent engineering staff. Company B obviously has smarter engineers.

One of the wonderful facts about the competition is that it is always ahead. But this is always only a temporary situation, which is about to be rectified by the current plan. So it is at our meeting. Instead of hanging his head in defeat, as would seem proper for a Company A engineer, one attendee raises his hand to voice his inspiring idea. "We will leapfrog them!" he proclaims with confidence. "What a marvelous idea," the managers echo. "Just what we were about to suggest ourselves." This is ingenious. While the competitors in Company B are stuck with a product that uses 37th generation technology, Company A will move immediately to 39th generation technology, thereby skipping a whole generation and moving decisively ahead of its competition.

How often have you heard someone suggest leapfrogging the competition? Has it ever truly worked? My experience has been dismal. The illusion, however, is comforting and inspiring. Generally, the problem is that leapfrog is a game for two. It requires a leapfrogger and a leapfrogee, and cooperation between the two. The leapfrogee must stand still and assume the proper position, while the leapfrogger makes a running jump. In the case of Company A, it will be extremely unlikely that Company B will be assuming the proper position. In fact, quite the opposite may be occurring. There is a meeting at Company B, decrying its deteriorating position with respect to Company A. An engineer at Company B raises his hand with an inspiring idea ....

Not only does leapfrogging fail on the grounds that Company B becomes a running target, but it fails because of the nature of technology itself. Very seldom is the 39th generation technology available to a company using the 37th generation. The 39th generation will be available in its own time, and to nearly all companies at the same time. Company B will also have 39th generation technology, but both companies must pass through the 38th generation. Like fine wine, no generation may be opened before its time.

Whether or not a leapfrog operation is under way, there comes a time in any project when a review of plans is made. The engineers in company A present the specifications of their advanced, intelligent widget to their management, comparing it in a most favorable way with the current product from the competition, Company B. For example, while Company B is stuck with an ancient 8088 processor, Company A's new product will use a powerful, cost-effective 80286. The battle is won! The management cheers the engineering triumph. No one seems to realize that at that very moment Company B engineers are presenting specifications to their management of a superadvanced, supremely intelligent widget using a 68020 processor. But somehow the plans for new products are always compared against existing products of the competition. Time after time we do this, knowing that it is not a fair comparison, until we believe it ourselves. "We're number one!" we cry out over our specifications. An impartial observer might mutter to himself, "Who do those turkeys think they are?"

The same ostrichlike behavior occurs in the critical business of choosing among competing technologies. An alternative memory technology is invented. The information in this form of memory is stored through the presence of vortices in turbulent gas; it is called tornado technology. A small group of researchers becomes interested in and eventually enthusiastic about tornadoes. Summoning a meeting of upper management, they cite the advantages that tornadoes will have, when suitably perfected, versus conventional memory technologies such as semiconductor and magnetic media. This presents management with a classic dilemma. Should it back tornadoes? The question is clouded by the knowledge that Company B already has a large task force assigned to tornadoes. Company B has, of course, taken this step because it has seen Company A's publications, which seem to indicate that a large initiative is under way.

At the meeting where the tornado decision is to be made, management listens to the tornado advocates cite their advantages for certain sizes and speeds of memory. There is no chance that a knowledgeable magnetic-media person will be present — someone who would be in a position to state boldly that improvements in magnetic recording will blow tornadoes out of the sky before they ever have a chance to ride the learning curve. The tornado people themselves have almost certainly underestimated the progress that will be made in competing technologies. This underestimation is not deliberate. It is simply the way things are. Tornado people are not supposed to be optimistic about magnetics. If magnetics were better, there would be no tornado people. This is an uncomfortable thought. Even though the tornado people think they have been fair to magnetics, they have not been optimistic, and history almost always shows that optimism is warranted in fields that are saturated with people and money.

Occasionally, of course, a new technology will overtake and outshine the best efforts of the competition. This is what makes the game interesting. It usually happens not because the competition fails to meet expectations, but because the new technology exceeds them. In the matter of the competition between companies, the truth is really very simple; whichever has produced the most recent product is ahead, by definition. I hesitate even to suggest this, but the relative brilliance of the engineering staffs may not matter so much. Technology moves so fast that a design produced six months later by an inferior engineering staff may be better than the original one produced by a brilliant staff. *Sic transit gloria* — the cry "We're number one!" passes, for half a generation, to other voices.

Proprietary Information

The Vugraph consisted of a single large word. "INTRODUCTION," it said. In small letters at the bottom of the foil was that ubiquitous warning, "Proprietary information, XYZ Corporation." I glanced around warily. Was there any unauthorized person present who might glimpse that secret word "introduction?" I had a twinge of the feeling I experience when I play a rented videotape and see the "FBI warning" notice that must be endured before the actual movie comes on. I always expect a loud knock on the door, and some tough-looking guy in a trenchcoat waiting there to arrest me for some inadvertent but illegal use of the rented tape.

The lawyers are nervous about engineers giving away precious information. They have a point. After all, XYZ Corporation has paid for the thought that went into the generation of the word "introduction." Seeing that word might save a competitor the development expense of coming up with their own word for an opening Vugraph.

What a different story it is at a professional engineering meeting, like an IEEE conference. There we find engineers showing block diagrams and algorithms, and discussing in detail all the reasoning that went into their products. Out of their pride in their technical accomplishments they babble on and on about the details. The speakers will tell you everything they know and then some. Of course, there are no lawyers around to stop them.

In contrast, the intellectual property people have this concept of a virtual great wall that can be drawn around their company, keeping all the inside information in, while allowing all the outside information to diffuse right through to the inside — sort of a lobster trap for information. No golden rule here, and no nonsense about being better to give than receive.

I try to imagine a world where that informational wall is strictly enforced. The IEEE journals and conferences would be banned. Instead, IEEE would become a kind of secret society. We would probably have to come up with a special handshake and some obscure Latin motto. Then when we met other engineers at social functions, we could identify each other through the handshake and the password. Of course, conversation would have to be guarded because of the no-fraternization rule.

"Oh, so you're an electric..., I mean, maybe we have something in common," you say, glancing around nervously.

"Uhmm ...," is the noncommittal reply.

"Uh, do you have a particular field?" you whisper quickly.

"Just ..., just general electricity," the other engineer confides helpfully.

"Well, is it, like, large electricity, or, say, small electricity?"

At this point a prosperous-looking man in a three-piece suit approaches. "Perhaps ... but I think it will rain tomorrow," you finish. No engineer would dare to look prosperous or wear a vest; this could be the intellectual property police.

Products would be designed in such a way as to protect trade secrets. All the important functionality would be encapsulated into very small, sealed plastic packages. Nobody would understand what was inside. Software would be written in abstract languages with huge amounts of extraneous, incomprehensible, diversionary code, much of which would be riddled with bugs. No product would work with any other product, unless it was from the same manufacturer. Standards and open architectures would be forbidden — a HAL computer would work only with HAL software. Corporate heaven would prevail.

Of course, engineering progress would be slowed to a creep in such an environment. The world would be deprived of advanced computer games, boom boxes, and stealth bombers. On the other hand, engineers would be guaranteed steady employment. The term "not invented here" would assume a new and frightening meaning, and every company would need its own stable of wizards.

An intellectual property expert reading this fanciful scenario would certainly be entitled to protest. "Let's be reasonable here," the expert would say. He would point out that in reality there are two kinds of information. There is green information, which can be shared with anyone, and red information, which needs to be kept proprietary. Green information has to do with theory and other mathematical stuff that is incomprehensible to lawyers, whereas red information is the real stuff — how to build a best-selling widget out of sand, that sort of thing.

The responsibility of the intellectual property people is to protect the company's own red information. The job of the industrial relations department is to uncover or steal other people's red information, while the engineers have the task of creating red information out of other people's green information. It's simple.

"Posh!" we would say to the so-called expert. Almost all information is actually yellow. Furthermore, many engineers feel that you have to give away some pink information just to balance your account in the international research community.

"Rubbish!" the savvy expert would say. Nobody keeps score. The journals and meetings are there for everybody, whether they contribute or not. Engineering knowledge is like a lighthouse, shining for one and all.

And so it goes. No one really understands how the world works, and perhaps the world itself is changing faster than we can understand it anyway. The thought even occurs to me that we might stop all meetings and publications, and then discover that no one had noticed or cared. Could we face the awful realization that no one actually does read the *Transactions*? I mean, we always joke about such things, but what if it were proved true? What if we discovered that information was really conveyed somewhere in the fourth dimension, that it seeps under windowsills and doorways, and that somehow every engineer just "knows" things without formally being told?

There are lots of fundamental mysteries here, but usually we just muddle through. As long as no one tries to really examine the process of information transfer, it seems to work fine. Up close, it just isn't very pretty.

Tahiti or O'Hare?

I have in front of me a travel request for someone to attend a technical conference in the French Riviera. Should I approve? How do you feel about such travel? For myself, my faculty for logical reasoning gets misted over with old memories of great and not-so-great meetings that happened once upon a time. To be honest, the first feeling that crosses my mind is jealousy. I imagine that while I am enmeshed in paperwork here in New Jersey, this person will be basking in the Mediterranean sun, all expenses paid by our benevolent employer. Why should I approve such a boondoggle?

Somehow in our engineering profession it is considered very bad form to admit to enjoying a conference. Ask any engineer returning from a meeting in Tahiti how it went. The schedule was murderous, you will hear. Sessions went on all night. Fortunately, you are told, through the diligence of this particular employee much useful information was obtained. Although the attendee is exhausted from the grueling travel, the conference was extremely worthwhile. You hear nary a word about the beaches, the restaurants, the sightseeing, or any of the images your mind conjures up at the mere mention of Tahiti.

Of course this hypocrisy comes back to bite us when we really do work hard at a conference in an exotic location. Just try to get some sympathy for the long, difficult hours you spent in darkened projection rooms in Bermuda. The dialog goes something like this:

Loop:
You: "No, I really did work hard."
Friend, Spouse, or Whoever: "I understand perfectly."
GOTO Loop

I suppose there is some kind of ethical mind-set hidden more or less deeply in all of us that feels that it is wrong to have too much fun at a technical conference. Meetings just should not be held in a place that is too ostentatiously nice. To appease this sense of conscience, we might consider holding all meetings at Chicago's O'Hare International Airport in the dead of winter. For those of us in the United States this would be a central location, and we would be unlikely to be accused of fostering boondoggles. I am just afraid that not too many people would show up, and that the mood of the attendees would not be conducive to the best of interchange. (The city of Chicago in the right seasons is quite another matter.)

When I consider my own reluctance to attend meetings in austere and lifeless locations, I come to the conclusion (or is it rationalization?) that as long as we can maintain professional demeanor, there really is no reason why a meeting should not be held in an agreeable location. Maintenance of this "professional demeanor" can be trying, of course — I have often been struck by a kind of catch-22 in conference attendance. It is almost universally recognized that in order to get permission or funds to attend a conference, one of the most important considerations is whether or not you are giving a paper. Yet I remember that in my early years I got less out of those conferences during which I gave talks than those where I did not.

If I had a talk to give, I was unable to concentrate much on those sessions that preceded my own because of apprehension about my own speech. After my talk, I felt relieved of all responsibility for concentration on what followed. I had earned my trip. If, however, I did not talk at a conference, I worked very hard throughout the conference to absorb as much as possible. After all, I had to earn my trip.

I remember great meetings, and I remember great boondoggles. Sometimes they were the same event. I remember the deer frolicking in the morning dew of a mountain meadow on a ranch near Santa Barbara, Calif., during a spring workshop. Everything came together that time. The harmony of the setting was a catalyst for genuinely meaningful technical interchange in the best spirit of our profession. On the other hand, I recall a conference held in a beachside hotel where no one showed up at a particular session except one speaker, who was dressed in a bathing suit as a precautionary measure. I am embarrassed to relate that incident, as it reflects poorly on our professionalism. I really think that the setting in that case was as much to blame as the participants, and furthermore that this fiasco represented a rare exception. Indeed, if this sort of behavior happened often, conferences would die away from nonsupport.

In the dear, dead days beyond recall, those of us from the United States felt that we resided in the center of technology. Since there was no real need to travel abroad, the notion of visiting other continents for technical meetings was associated with a perk or reward. Now we keep hearing — lest we forget — about the emergence of the so-called global economy. Perhaps in no field of endeavor is this globalization so pronounced as in engineering. Chicago may be the geographical center of U.S. technology, but it is not necessarily central to the global diffusion of excellence.

Centers of excellence in technology exist in many different countries, and conferences represent one of the most important means of promoting the flow of technology across these international boundaries. I think of conferences as the nodes in the informal communications network that binds our profession. Not only is information exchanged, but friendships are formed, establishing links that long survive the transient duration of the conference. Given the nodal importance of meetings, and the global nature of our work, it is inescapable that we in the United States should travel more abroad.

Having said that, I look at the approval request in front of me and admit that my biases — like those of many of my peers — were formed in those years when foreign travel was a perquisite of the rich, the famous, and the fortunate. (Sigh.)

Perhaps even more important than my biases is the fact that my foreign travel budget is historically derived from that yesteryear of imagined dominance. To compensate, someone has suggested to me that the foreign travel budget should be indexed to the "balance of trade" in technological information. As this balance moves abroad from the United States, the money spent on foreign travel by U.S. engineers should be proportionately increased. This makes a great deal of sense, though I despair of explaining it to the accountants. Because of the reluctance of business and government in the United States to fund foreign travel, it is probable that IEEE schedules fewer conferences in non-U.S. locations than would be warranted by the distribution of engineering excellence.

Now that I have carefully and conscientiously thought over the matter, I have just signed the approval for the trip to the French Riviera. I'm still jealous.

Committees and Canines

In one of those famous *New Yorker* cartoons by Charles Addams, a man and a small boy walk down the sidewalk in front of imposing, governmental buildings. A number of large statues dominate the foreground. In each one, instead of a heroic figure on horseback, a group of bureaucratic wimps is portrayed. The man is saying to the child, "There are no great men, my boy — only great committees." Only a cartoon could convey so well the contradiction in terms. The very idea of a statue of a committee brings a smile to our faces; yet engineering, the IEEE, and life-as-we-know-it are filled with committees on which all of us serve with a shared pride in our accomplishment.

In the IEEE, of course, committees are endemic. (Not that the IEEE is to be confused with real life.) Some years ago random events and the turn of dice brought me to the presidency of an unnamed IEEE entity. Here was my opportunity to salvage a rotting bureaucracy. Committees, subcommittees, and such abounded for who knows what long-forgotten purposes. Mostly, they just abounded.

Seizing the management initiative with my newfound executive power I eliminated the more obscure of these wasteful legacies. No sooner was the word put out through the IEEE town criers than my phone started ringing with calls from the bewildered, disenfranchised ex-committee members. How could I do such a thing? What would they tell their companies and universities? How would they be able to attend conferences in attractive faraway places? How would they be able to meet with their friends? Worst of all, had I no compassion for the terrible deflation of their resumes? Had I no pity whatsoever?

Needless to say, I relented. Subsequently I came to an understanding with myself that was either rationalization or keen insight, depending on your own viewpoint. The IEEE is not a corporation, meant to be run efficiently with brave, crisp executive decisions. Every volunteer, with good reason, wants a real voice in decision making. After all, the IEEE belongs to the volunteers, who give enthusiastically of their time for the cause of our profession. Moreover, labor is free (at least to the IEEE); thus efficiency is not so important. The IEEE has never to my knowledge had a layoff of volunteers. Why not have all sorts of committees?

I can of course think of many committees that have accomplished real work. Standards, for example, is an area in which committees have occasionally produced consensus work that has had an impact on our profession in a beneficial, and sometimes even insightful, way. Theorists who argue about the relative virtues of man and machine point out the ability of people to work together to produce reasoning and results superior to that of a single person. Although it does not seem impossible for machines to emulate this behavior, it certainly appears a far-flung dream for today's computers.

I think all of us have been part of a committee where we experienced a kind of group creativity, where ideas were so amplified and reinforced that no one even recalled from whence they came. Our profession today is so broad and complex that, although there are more than a million electrical engineers, each of us is undoubtedly professionally unique. More than ever, we need mechanisms like committees to enable us to solve problems bigger than ourselves.

On the other hand, we have all belonged to the other kind of committee also — the camel-designing committee. I would not say that we engineers specialize in this kind, but we can certainly hold our own with about anybody. I recall some years ago being given (a strange word, "given") the chairmanship of a particular IEEE committee. The task was controversial, the assignment murky and difficult. (To those of you who know what this committee was, you are wrong.) After a year of meaningless, rhetoric-filled meetings I could no longer face the embarrassment of calling together busy people from across the country to accomplish so little.

I gave a final report to the board that had appointed us, saying in essence that we had failed in the assignment. Much to my surprise, there was applause and a motion of thanks for a job well done. Graciously, I accepted the congratulations as I slowly came to understand the situation. No one had really wanted anything done by the committee in the first place. *They* had done *their* job by appointing me chairman. My only real failure had been in not appointing a subcommittee to do the work. Then I could have passed a resolution of congratulations, and the subcommittee chairperson could have been left wondering why he or she had failed to appoint a sub-subcommittee.

Committees develop personalities all their own. Sometimes I liken them to dogs. There is the purebred committee of the poodle variety — not very useful, but its nose is held high in the air. I always wanted to be appointed to such a prestigious committee, where members collect much fame and power while doing essentially no work, but every time I join a committee and look around, there are only people like myself. Why is that?

The other day I had my chance. The chairman, in extending the precious invitation to join his blue-ribbon committee, told me of several famous people who had already agreed to be members. At the first meeting of the committee, guess which members were absent? It is possible, I am told, that these famous people will actually read the final report of the committee. It is not certain, however.

Then there is the mongrel committee, which must scrounge for its very existence. Usually this is called a "working" committee, to distinguish it from the one that appointed it, which is a "nonworking" committee. The members of the mongrel committee are a little of this and a little of that. They don't pull together very well, but manage to tip over a lot of garbage cans and pick a lot of fights. There are terrier committees that frisk about nipping at everyone's heels, creating the illusion of progress without any coherent purpose whatever. And there are bulldog committees, single-minded and relentless. Some committees are giant and hairy, filled with inertia and lethargy, while others are lean and wiry, with boundless energy and innate hunger.

The beauty of it is that one never knows in advance how a given committee will turn out. The committee's individual chemistry is a complex function of the chairperson, the members, the task, and the phase of the moon — mostly the latter.

Is there a better way to run a professional society? I'm not sure, but the next time I am appointed to a shaggy dog committee, I'll give it some thought.

Turf

In *The Territorial Imperative* Robert Ardrey discusses the curious propensity of animals to regard areas of space as exclusive possessions and their compulsion to defend these areas against all intruders of their kind. The male gibbon, for example, trumpets a great siren call, which is an announcement of his location to his neighbors. This serves as a warning to all that trespassers will not be tolerated, and when the gibbon is in a restless mood it is an invitation to any like-spirited gibbon to appear on the boundary to do battle. Ardrey asks why — why spend so much energy in the defense of a portion of land indistinguishable from the next portion? Clearly gibbons are irrational creatures; someone should tell them that their behavior is silly. Fortunately, electrical engineers are much more intelligent than apes. You would never hear an electrical engineer emitting a ridiculous siren call.

Suppose that there were a planet inhabited by apes, some of whom were, of course, electrical engineers. Let's call the ones we're talking about gorillas. On such a planet our profession would be in bad shape. The gorilla engineers would have an exaggerated sense of territory, or what we call turf. I can imagine a department engaged in the design of a mobile phone system, probably using a cellular concept to track customers from tree to tree.

Naturally this department would feel that its responsibility encompassed the entire system. In the design of the critical ape-machine interface, they would not feel it necessary to check with the experts in the simian-factors research group. No, they would want to do it their own way, unconstrained by irrelevant advice from apes who do not understand the overall system context in which such an interface would be embedded.

Probably word of this slight would reach the researchers by and by. Perhaps some careless grunts would be overheard during a banana break, or the message would spread through the company monkey vine. On such a planet we would probably hear the research gorillas trumpet their great siren calls of territorial claim. The trumpeting research gorillas would look on with disdain as the development gorillas proceeded with their design. More and more they would be heard to downgrunt the approach being taken. They would absolve themselves of the inevitable market failure that would ensue. This is the way apes are.

Meanwhile I can imagine other trouble on this ill-fated planet. While the mobile phone system was being developed, there would probably be another group in the company charged with the development of the cordless phone. Such a phone would be a sure market success as it would eliminate the frustrating wires that would be forever getting tangled in the branches of the typical home. The cordless group would assume that such a phone would be its exclusive territory.

The onset of the development would probably be accompanied by a series of annoying siren calls. Although there would undoubtedly be considerable overlapping in purpose and function between this phone and the phone for the mobile system, neither group would consult with the other group. As a result the evolving terminals would be incompatible. Upon their discovery of this debacle, there would be a lot of chest-beating and paw-pointing among the management gorillas. They would suspect the worst; after all, who knows what evil lurks in the hearts of apes? This is the way they are.

A series of meetings would ensue. Each development team would claim that the requirements for its project were so unique as to compel a dedicated solution. No input would be necessary or desirable from the other group. The gorillas would all look innocent as they absently scratched their sides and stared off into space. There would, of course, be excellent technical arguments showing why the projects had not been coordinated. No gorilla present would be so discourteous as to point out the simple fact that the two teams inhabited different territories. The management gorillas would try to push the incompatible projects together, since this would be the time-honored, suicidal mission they would have had pressed upon them by their ancestors. Perhaps no gorilla would even remember why this was a good thing to do, but this is the way apes operate.

Knowing apes, I would guess there would be other complications. It is likely that there would be a marketing department full of young, ambitious orangutans. They would live in an entirely separate jungle and probably have their own silly little territorial instincts. They would busy themselves segmenting the market and doing surveys of the average tree owner. While they were busy compiling a list of desirable features, the engineer gorillas would already be approaching final design. A confrontation would be unavoidable. Probably some feature that the marketing department felt a necessity would be impossible to incorporate. The orangutans would make silly barking noises about how it was supposed to be a market-driven company. The engineer gorillas would chide them, then they would roll their eyes and suppress yawns. What can you expect from an orangutan? they would ask. The orangutans would gesticulate rapidly to each other. That's the way gorillas are, they would say.

If we could visit a planet like this we would have much to learn. But perhaps we would see communications satellites and optical fibers spanning the jungles, microprocessors and mainframes, very large-scale integration of incredible complexity, and dazzling technology everywhere. Maybe this, too, is the way apes are. We might wonder if there is any connection between this massive innovation and the territorial trumpeting. Is the one necessary for the other? Or are these apes just as silly as they seem with their ridiculous concerns over little pieces of turf that nobody else really cares a siren about?

That Dreaded NIH Syndrome

A pollster stands on a city street corner asking passers-by what "NIH" means to them. Most people shrug their shoulders and hurry on. Occasionally someone presses a quarter into the pollster's hand. Some of the better-dressed answer "National Institutes of Health." But just every now and then someone glances around with a look of guilt and, with eyes fixed on his shoes, mutters under his breath, "Not Invented Here." "What's that again?" the pollster asks. "I said I don't know," is the embarrassed reply as the person rushes off. An engineer has been identified.

An engineer aspires to invent. The fact that other people run around inventing things greatly complicates this noble aspiration. To combat this unseemly behavior in our fellow engineers we have evolved a protective mechanism consisting of a whole arsenal of weapons that can be used to deny these "foreign" pretenders posing as legitimate inventors. This dreaded NIH factor apparently comes to us in an instinctual fashion. The last I looked, admittedly some time ago, there were no college courses entitled "Elements of Invention Rejection I" or "Advanced Maligning of Inventions in (say) Power Transmission." Nevertheless, somehow we hear the drumbeats in the ether of our professional environment. "It can't bode well for me if this other person's idea is a good one," we think. (And having things "bode well" is, after all, a time-honored objective.) Of course we don't really think this subversive thought with our real brain. I would never think something like that, we say to ourselves, my obligation is to truth (and justice for all) and dispassionate consideration of all competing ideas. We can't help it if these ideas are so much inferior to what we happen to be working on ourselves.

The first line of defense is to point out that the offending idea is really an old one — preferably a very old one. This classic defensive maneuver has the advantage of demonstrating your superior knowledge of the literature. "Your idea for an airplane is certainly a good one, Mr. Wright," you say, "but before you spread this around too much you should know that this concept was published quite a few years ago in a very generalized fashion by a Mr. Da Vinci." You add modestly, "I've read the paper in the original Italian. Perhaps I can have my secretary make a copy for you." When in doubt, you pause and observe hesitantly that you saw something like this in a paper by Pliny the Elder many years ago. Generally speaking, you are well covered on this one; the haziness of your memory and the extent and ambiguity of the literature will keep the poor impostor occupied for months.

If you are unable on the spur of the moment to dismiss the alleged invention as old hat ("Well, Mr. Stetson, this plan you have for a thing to wear on one's head ..."), you're forced back to the generalized "can't be done" argument. Here a subtle ridicule is especially effective. "Now let me get this straight, Mr. Sarnoff," you say. "I think that I understand how it's at least theoretically possible to transmit the color information in the way you suggest. But supposing that we could actually accomplish this, you're saying we could put this ... this shadow mask inside the tube to separate the colors." (Here you partially suppress a giggle.) "Now you're suggesting that we could build this receiver for several hundred dollars and put one in every home. For an instant there I thought you were actually being serious, but I'm so naive that I don't even realize when my leg is being pulled."

When other defenses have been exhausted, you can always pull out the "nice, but useless" dismissal. Your dialogue can be modeled on the following: "This thing you call a transistor is really rather cute, Dr. Shockley. I can even foresee the day when we could make these into tieclips as little mementos of the curious byroads that physics research can take. Frankly, though, it's hard to imagine that there could be any application to actual electronic design. I know you're thinking that the small size could be an advantage, but you probably don't realize that these things would have to be mounted on octal sockets. There would have to be some kind of enclosing envelope that would be large enough for you to get your fingers around for insertion. So you see it would actually be the same size as a vacuum tube, and I don't want to disillusion you, but you should see the new miniature tubes that my group has been working on. Then, too, not being accustomed to electronic design, as I am, you might think that not having to heat a filament is another advantage. But you can be excused for not knowing that all transformers are constructed with 6.3-volt windings just for this purpose. I certainly would think that this work could lead to some good theoretical papers, but you should be careful about talking too naively about this around electronic people who may not be the good friend that I am."

Sometimes — just sometimes — we're on the other side of this NIH barrier. We get this great idea for a better widget. Gathering our papers in great haste, we race over to the widget-development department to tell them of their salvation. Right off, we get referred to the Pliny the Elder reference. You'd think, having used this strategem so often ourselves, that we would recognize its hollowness. But no, if you're like me, you beat a hasty retreat, trying awkwardly to cover your humiliation at having been so thoroughly beaten out by Pliny. You run to the library and search with a combination of desperation and furtiveness (should this be possible) for the Pliny reference. The reference trembles in your sweaty hand and blurs before your vision. After the first few seconds, when it looks like all is lost as you see familiar figures and equations, you begin to spot the differences. It's as if you passed your file and the Pliny file through a differencing filter. Only the points of difference print out on the CRT in your mind. Pliny didn't do anything like what you did! After all, the phased-locked loop in Pliny's reference is analog, whereas yours is digital. With great indignation you ....

What is it that you do? Does the dreaded NIH syndrome win again? Or are the widget-development people right? Such is the everyday soap opera of our profession.

Finger-Pointing

Have you ever been associated with a failed project? Probably you haven't; no one else I know has either. All the biographies I read speak of successes. Failure does not occur. Yet somehow I remember various schedules that weren't met, programs that didn't function, research that fizzled, and products that didn't sell. It's funny — all the people are successful; it is only the projects that fail.

There are some basic laws of life involved here. We all know that things will go wrong, as is well documented in the classic paper by Murphy. With electrical engineering projects as complicated as they are these days, there is no question that things will inevitably become a muddle. The only question is — who will be blamed?

It seems that we all have an innate ability to escape guilt. Either that, or finger-pointing is the immediate subject of on-the-job training, because little of our educational experiences seem relevant. When you fail a test in college, your name is at the top of the paper, next to the "F"; pointing your finger at other people is not effective. In the complexities of the real world there is no "F" at the top of the paper, but — more important — there is no name there either.

Why is there no name to be blamed at the top of a real-life test? Most companies talk tough about rewards for success and penalties for failure. It just seems that in the latter case no one can be found to assume responsibility. (In the case of success, there does not seem to be a problem finding a plethora of responsible parties.)

In the event of failure, there are often organizational reasons why no guilty person can be found. In the turmoil of modern life in the high-tech world, reorganizations are so frequent that it is not clear in retrospect who was in charge at the time the failure occurred, since this instant cannot be properly isolated. It would be a simplification to say this is exactly like a game of musical chairs where someone is left standing at the moment the music stops. Picture instead the music fading away gradually and the cast of players constantly changing in confusion, while a maintenance crew is mysteriously removing chairs in the midst of the competition.

Have you ever noticed that some executives seem to leave behind a wake of disastrous projects, always being promoted out of a job before it sours? They can always insist the project would have been a success had they remained in charge. Timing is everything, but even extraordinarily bad timing can be overcome if the management level is high enough. For example, I have even heard a theory of promotional advancement that depends upon being at the helm of a debacle when it crashes. Naturally one's career undergoes a momentary lull, but in later years people forget the history of the unfortunate project and remember only the name of the manager. "Oh, of course I remember Stan," they say. "We'd be lucky to have someone as well known as he to accept this modest promotion."

At the bottom of the organizational structure, the tactics for guilt avoidance are necessarily different. Cunning and craftiness are often needed to supplement questionable engineering. There is a strategy usually referred to only by its acronym, CYA, that many believe to be effective. For example, before a final design is submitted, you issue a memorandum pointing out possible weaknesses in the data and components supplied by other company divisions. Of course, it will be necessary to defend your own design against their allegations, but always leave yourself an out. Your design utilizes the finest technology and the latest principles, but you are depending on the integrated widget being available on schedule and meeting certain specifications. You would not recommend your backup design. (Don't worry about the integrated-widget people becoming the villains should things go awry — they are depending on the device department, which in turn has been promised certain materials from a subcontractor.)

I wonder how George Washington would handle all this if he were an electrical engineer? Chopping down cherry trees isn't in vogue the way it used to be. If George were a designer, he would be anxious to admit to any error that he had committed. "I cannot tell a lie," he would tell his associates. "It was my design of the bus interface board which caused our product to fail."

His friends would give him wise counsel, both because George looks like an up-and-coming engineer and because guilt in an organization can bring lightning from the corporate sky. "George, you're not being modern," they would say. "How could you have known that the systems department had underspecified the drive capability your board would need? The marketing people added features at the last moment. It has nothing to do with us, George."

I don't think George would have been dissuaded. Believing in the domino theory of "for want of a nail, etc.," George would go to his department head and confess. "I cannot tell a lie," he would repeat. "It was my design that caused our company's quarterly return to be below expectations."

At the next board meeting the president would be fuming about the dismal performance in the market. "Heads will roll for this," he would threaten. His underlings would jump to the defenses. "We have already conducted an investigation, and we have identified the person responsible. His name is George Washington. He works in the widget design department." I can imagine the president becoming very irritated at this disclosure. "I've never heard of George Washington," he might shout. "I want to fire someone important! What is the name of Washington's department head?"

In the contemporary engineering world, guilt is a very abstract concept. When I think of my George Washington fantasy, I realize how difficult it would be even to claim guilt. So why do I see all the finger-pointing and people covering themselves against contingencies?

Perhaps it is all a game we see played out elsewhere in the business world. Somehow, I'd like to think engineers are just a little above all that. Maybe tomorrow George Washington will come into my office with his confession, and I'll see how well I fare with my own convictions.

Part 5

Looking at Life

Sorry for the pretentious title of this section. Engineers have to live life the same as everyone else, and occasionally my essays have had more to do with ordinary life than with our profession. Even so, it is ordinary life as seen through the eyes of an engineer, which means there is a certain color cast upon everything. It is a kind of worrisome color, maybe indigo. Who else would know about indigo?

There is an expression about wearing your heart on your sleeve. In the case of engineers and scientists, it is often the delicate ego that is worn on the sleeve. I can find this theme in a number of the essays in this section. A shrink would probably find something very significant about all of this. Just lie down here and tell me about it, he would say to us engineers and scientists. (You don't think I am taking the blame alone, do you?)

Some of my favorite essays are here in this section, and I would like to start with the one I like best, "Geniuses." On several occasions I have seen this essay pinned to a wall when I was visiting some unknown company, and I have felt a special kinship with whoever put it there. It also has made me feel vulnerable, because this essay comes too close to truthfulness to be comfortable. It is about the inadequacies that we all secretly feel that we have when compared with so many of the smart people we see competing against us out there.

Most of us have known people in our childhood that everyone has said were geniuses. There is nothing quite like having someone tell you that someone else is a genius to put you in your own place. A friend comes up to you and, lowering his voice, begins, "Just between us ordinary people ...." This has to give you ambivalent feelings. In school it is imperative to be considered ordinary in order to be accepted into the most desirable communities at that age. All the same, being lumped with all those ordinary people, especially by someone who as a close friend knows you well, has to leave little scars in the psyche.

Many, if not most, of us did well in school and yet few of us had the wisdom at that age to see ourselves through the eyes of others. A very, very bright person that I know was recently reminiscing about his high school math courses. "To the other students I was the *enemy*," he reflected. I imagined him in my high school math course. Yes, I thought to myself, he would definitely be the enemy. We all had them in those days. The first day in a class you could look around and count the enemies. But I believe that few of us saw ourselves as the enemy at that time. For all we know now, people pointed at us and whispered to their friends that there goes so-and-so, the genius. You, perhaps.

Isaac Asimov, the celebrated scientific writer, once gave me his business card. It was the simplest card I have ever seen, and the most impressive. It looked like this:

Isaac Asimov
Genius

Maybe Asimov was a genius. I would give him the benefit of the doubt, but I never knew him personally. I have known a lot of people that are considered geniuses by those who do not know them, and none of them has lived up to the title in person. Leaders of government and industry, generals, and the world's greatest scientists and engineers have all seemed ordinary to me on occasion. They make the same mistakes and have the same uncertainties as you and I. Well, me for sure. I can never be sure about you. Maybe you are the elusive genius.

The lesson in life that I discuss in "Geniuses" is one that I continue to receive. A few years ago I heard a talk at a military briefing about some great new discovery. I really did not understand very much of it, but as I looked at the famous people around the table, I told myself that I had better keep my ignorance to myself. Obviously, these people were all following the talk. Probably they were way ahead of the speaker.

When the speaker finished the talk, he stared at the audience. "Any questions?" he asked. Silence. Silence. Then one hand reached up. It was not just any hand, but the hand of, arguably, the most famous physicist in the world. The quiet deepened. Everyone strained to hear the famous voice, and the surely profound question that would be posed. The great man shook his heavy jowls in an expression of disgust. "I have ... I have understood *nothing* of this," he spluttered.

Time and again I have learned. And time and again I have forgotten. Chances are that if I do not understand something that is being explained to the nonspecialist, then not too many other people will understand either. The truly great people are the ones who can admit their ignorance in these circumstances. How many of us could say that we have "understood nothing" of some simple talk? Yet I cannot recommend that you adopt the strategy of making these sorts of confessions. I said that it takes a great person to make such statements, but the nub of the matter is that it *requires* a great person. The famous physicist that I mentioned was not really confessing a personal inadequacy, rather he was making an implicit criticism of the talk. Unfortunately, if either you or I made such a statement, it would be taken as a personal failing, and the others in the room would keep quiet about their own states of understanding. It is that kind of behavior that reinforces our suspicions that other people may be smarter than we are.

Amateurs, Experts, and Dinosaurs

The perceived expertise and intelligence of others is a theme that appears in several more of my essays. "Diminishing Dinosaurs" was originally entitled "Amateurs." It is about the psychological barriers that keep us from messing about in matters in which we are not expert. Of course, the experts in whatever field it is would like to keep it exactly that way. This is the most important job of being an expert. You must defend your territory against all comers. Naturally, this defense is made easier if there are fewer comers.

I imagine some course in the college catalog entitled, "Elements of Being an Expert 101." There are many strategies that must be mastered. First and foremost is vocabulary. Preferably, no one untrained in the speciality should be able to think or communicate an idea in the field, since they will not know what words to use or how to use them. Take away the words and what do you have? For example, you never hear a doctor saying silly things like the kneebone being connected to the thighbone. That would make it understandable. Moreover, experts can have conversations among themselves that are entirely incomprehensible to a bystander, while at the same time sounding enormously impressive. In overhearing the Latin terms spoken by communicating physicians, all you can think is that this is why your medical bills must be so high.

After vocabulary, there are other defense mechanisms. Licensing or some method of restricting entry to the profession is often a fallback measure. If moats were still in style, even they would be used. Man the trenches; keep the infidels from the barriers! So when you are an infidel approaching the barrier, you are naturally reticent. Obviously, the people standing on top of those barricades know so much more than you do. There are two dual lessons in life that apply here. One is that these people may not know as much as you think they do. However, the corollary is that when you are atop the barricades yourself, secure as an expert, the amateurs below may know something you do not. Beware!

The title of the "dinosaur" essay is taken from a small, but memorable, event that happened to me at London's Museum of Natural History. Coincidentally, I have since had occasion to give talks both in that museum and at the other dinosaur hall that I mention, which is in the Carnegie Museum in Pittsburgh. I feel like these dinosaurs follow me around. I am reminded of a cartoon that someone sent me when the Bell System underwent divestiture in 1984. In the cartoon, taken from Gary Larson's *The Far Side*, there is a conference of dinosaurs taking place. One dinosaur is delivering a lecture from the podium to a crammed room of other dinosaurs. In the version I received, the dinosaur on the podium was wearing an AT&T badge. In its lecture, it was saying, "The picture's pretty bleak, gentlemen. The world's climates are changing, the mammals are taking over, and we all have a brain about the size of a walnut."

While I am on the subject of dinosaurs, I am also reminded of the inflatable Godzilla that was sitting behind my desk when I came to work on the morning of my 25th anniversary at my company. It was about seven feet high, green, and gruesome. For the morning I left it at my desk, while I worked at a side chair, amused by the expressions on the faces of people who came into my office and spied the Godzilla where I should have been. However, as the afternoon began, I needed my desk back. The Godzilla had to go, but what to do with it?

Here is where I had a minor brainstorm (very minor, actually). I pushed the Godzilla out into the hallway, two secretaries running along behind as I headed toward the elevator tower. When an elevator showed up at my floor, the topmost, I shoved the Godzilla onto the elevator, pushed the basement button, and got off. The Godzilla started its lonely trip to the basement in the empty elevator. I started running down the staircase, but the secretaries had a head start, and it was soon apparent to me that whereas they might make the basement, I had no chance to be there when the elevator doors opened to the surprised group waiting in the basement.

I settled for the third floor, where I pushed the "up" button. I had a one in four chance of getting "my" elevator on the way up. However, this time I was fortunate enough to have the same elevator stop for me. I wish I had a photo of what I saw when its doors opened. There was the godzilla standing serenely in the middle of the crowded elevator. The people were crushed up against each other in order to leave room for the big green monster. Not a single person had a smile on their face, or seemed at all interested in the fact that the principal occupant of their elevator was a seven-foot Godzilla. I don't know what that says about today's corporate environment, but it can't be good.

The dinosaur story in my essay is about how large the world looks when we are small, and how it shrinks as we grow. But all of us have these remnants of childhood locked in the chemistry of our brains, where there are all these people out there that know so much more than we do about everything, and we are discouraged from wandering very far from our own secure backyards of expertise. This applies to engineers as well as to everyone else, and in our case it is one of the factors that has made us reticent to enter the governmental fray and to deal with segments of the outside world that influence our destiny.

The IEEE was late, relative to other professions, in getting into political activities, but I believe that we have done a commendable job in recent years. Most members of Congress have at least heard of the IEEE now. They also have heard of a lot of small labor unions who may not always have been as bashful as we have been.

Keeping Score

Sometimes I reflect at the end of the day about how things have gone. Did I do well, or did I do poorly? Often it is hard to tell, but usually I have a feeling one way or the other. I see a giant scoreboard in the sky, and my score is either mounting or declining. Probably what matters most is the direction of motion; life seems to be a matter of the derivative, rather than the actual value of success or satisfaction.

Perhaps this feeling of scoring is intrinsic to keeping oneself afloat in the sea of modern life, but possibly it has its origins in our schooling, where we run the gauntlet of test after test. Today my calendar is full of meetings, appointments, talks, and various disagreeable commitments. My life at present seems to have been scheduled away beyond my control. Back when I was in college I did not even have a calendar. The world wasn't like that. What I remember, though, was that the entire sense of time was dominated by the succession of tests in the various courses that I was taking. I did not have to write down these dates; they were wired into my brain.

Those tests are still wired into my brain. In "The Dreaded Test" I wrote about a dream that I have had periodically throughout my adult life, in which I am supposed to take a test somewhere in some surrealistic college, and I cannot find the room where it is to be, or else I have never attended a single class in that course. I do not think I have any other such recurrent dream. Long ago I thought perhaps I was the only one haunted in this way, but I read that a poll of some college class at, I believe, Harvard, uncovered the fact that this is a very common dream.

A number of people wrote me after the publication of this article to recount their own variations of this dream. One that was especially poignant was an engineer who was at Pearl Harbor when it was attacked by the Japanese in 1941. He remembers vividly the death and destruction that were rained from the skies. But he said that he never dreams of Pearl Harbor — only of the tests in college.

The tests may haunt people like me, but I also remember college as a clean, compartmentalized existence. Reality was a series of textbooks, a sequence of homework problems. Every problem had one, definitive answer, right or wrong, and nothing in between. A test often meant an all-nighter, but when it was over, it was over. You could wipe all that superfluous knowledge right out of your head, so as to be ready for the next test. You knew instinctively that you would never need that information again.

When you finished a course in college, you got a definitive grade, and it was finished forever. You began again with a clean slate. During the summer you could wipe even the existence of college from your mind and take a sabbatical to do something totally different. If you worked during the summer and did not especially enjoy the work, it did not matter. You were a college student. Nothing was forever, and in the fall you would start anew.

Today I often visit college campuses. How I envy the students! If only we could all spend our lives as students! I see the students lying in the thick, green grass on a perfect spring afternoon, watching the frisbees sailing overhead — fleeting, white saucers in the deep blue sky. I walk by in my uncomfortable business suit, tortured by the tie choking my neck, and thinking how little of my time is my own. Though I am absorbed in the vision they present, my business suit and age make me completely invisible to them.

I see the textbooks lying by their sides, and the titles are often subjects about which I secretly wish I knew more. Someday I will learn more about that, I say to myself. But there is no time, and my promise echoes falsely even to myself. I understand the motivation of the graduate students who prolong their studies. Perhaps they have some inkling of what lies in store, of the workdays ahead for which their studies are a prelude, and they tarry to enjoy life while they can.

Yes, college always seems idyllic to me in retrospect. At the time I didn't realize how good it was, even though those tests, projects, and term papers hung over my head. Now, years later, I only remember the good times. That is during the daylight. Sometimes in my dreams, the tests return. I guess they never left.

Heroes

We all need heroes, and our profession has plenty to offer. Plenty, that is, in the past. The British IEE has a magnificent statue in the foyer of their building on the Thames. It makes me proud to be associated with the electrical field. Every now and then we should take someone, wrap his or her likeness in bronze, and throw the contraption on a pedestal. As an example that provides real inspiration, you must see the sculpture of Albert Einstein in the garden of the National Academy in Washington, D.C. I remember being skeptical of the cost when the Academy was raising money for this, but it is truly moving. See it, and you will understand.

In "The Footsteps of Giants" I worry about whether there will be enough contemporary heroes to go around in the future. On the scale of worries, of course, the absence of an adequate number of heroes would not be at the top of the list, but if it is true that heroes are disappearing, it says something worrisome about what is happening in our world. Today it seems more and more that projects are executed by teams, and the trend to software and virtual concepts deprives us of the artifacts that have been associated with great individual inventors. Perhaps in the future the statue in the entrance hallway of the electrical society will be a giant bronze blob seated in front of a computer screen. The title will be "The Unknown Engineer."

I have occasionally had the privilege of sitting on various awards committees. Perhaps my memory clouds, but in the early years the giants were plentiful. Dossiers were filled with brilliant inventions, hundreds of papers, dozens of books. It seems much harder today. The chief requirement to win an award is to have won another award. The other award becomes the outstanding accomplishment. You have to work your way up in awards, beginning with a small, starter award. Heroes have to be manufactured.

When I first went to Bell Labs, in the early 1960s, I was intimidated and awestruck by the famous people who drifted by the hallways. Now that the years have flown, I see the hallways filled with hurrying strangers in suits, looking suspiciously like business types. Perhaps they are engineers in disguise. Perhaps not. There is no question in my mind that things are different. Is it the world, is it us, or is it technology?

After the publication of "The Footsteps of Giants" I received a letter from one of those giants, Richard W. Hamming. Dick Hamming, in case he needs an introduction, invented the first error correcting codes, the Hamming codes, conceived the Hamming window for signal processing (nice to have things named after you!), and was one of the pioneers of computer science. His book in that field opens with his famous dictum "the purpose of computing is insight, not numbers." Come to think of it, who else has a "dictum" attributed to them?

Hamming's letter itself offers some of that insight about the phenomenon of giants as seen from time afar. It is worth reproducing in its entirety.

In answer to your "Reflections: The Footsteps of Giants," I too have have had a lot of time to think of what I lived through at Bell Labs during my 30 years, 1946-76, which is the time you are talking about. I claim that the period was unique and it is not likely to be duplicated. I was one of the four "Young Turks," and it appears both that we saw ourselves that way, and that management independently saw us that way. The four were: Shannon, Ling, McMillan, and myself, the middle two having risen to vice-presidents. That leaves me! We were all very close in age and hence had a common background from our society. Let me document the features that seem to me to have mattered.

1. *We were all of an age to be deeply affected by the Great Depression, and as a result felt that we owed the world a living, and not that we had only to behave ourselves and the world would be obliged to take care of us. This is a fundamental attitude that explains a lot of our drive to succeed.*
2. *We had all of us been forced (quite willingly!) during the war to engage in activities that were not what we would have chosen in normal times, and we were forced to learn a lot of strange, new things in science that were unrelated to our past training. We were broadened in our outlooks.*
3. *We had seen the almost immediate consequences of advances in science and engineering that we had been involved in, and as a result, we had greater reason to believe that striving to discover new things was worthwhile.*
4. *Our management had also been through somewhat similar experiences, and probably were more sympathetic to our unconventional behavior than would otherwise be true. Certainly they deserve much credit for letting us do the things we did the way we did them!*
5. *We had seen comparatively young people direct important work and bear responsibilities, and some of us had indeed risen early to positions of responsibility.*
6. *Due to the tremendous war effort (and we felt that unlike in any other war America was directly threatened in WWII) science had made remarkable strides, and there was a widespread recognition that science was important — more so than now. It was an exciting time to be a scientist or engineer.*
7. *Perhaps the coming of the computer should also be regarded a unique stimulus to creative science.*

My conclusion is that you cannot again create the situation that gave rise to that remarkable outburst of creativity.

The next group of people in the department that were hired were perhaps more able, better educated, etc., but collectively they simply did not equal our productivity. The average of about seven years in age apparently made a great deal of difference, but also the opportunities we had were to some extent denied to them.

Yes, it was a unique time — I was lucky!

Ordinary People

So much for giants. The complementary set is the rest of us. One theory that I sometimes harbor is that we all have little seeds of giants — Jack and the Beanstalk sorts of seeds — buried within ourselves. All we need is the fertile conditions for our intrinsic abilities to burst forward, but often those conditions never happen. Waiting for that season of brilliance and necessity, we simply coast through life. That is, sort of, what "Faking It" is about.

To be honest, I do not feel good about this essay. Only my sense of orderliness and completeness forces me to include it in this book. Sometimes I get trapped within an essay that I am writing, and it takes awkward turns. Afterward I look at what I have done and I do not feel right about it. Even though this one has a few nice stories and embryonic ideas, I am in some sense ashamed of it. How could I admit that sometimes I am a fake?

Yet the truth is that no one operates "full out" all the time. The most physically tiring days that I experience are days when I have to be in some sense on stage continuously. I go to visit some company or university, and in order to be good hosts, they always feel that they must fill up my day completely. Thus they usually put together a schedule of back-to-back interviews with everyone there. Every half hour a fresh person stares at me and expects to listen to some revealing wisdom. Even if they are doing the talking, listening in rapt attention is hard work for me. At the end of the day I am exhausted.

Sometimes in a meeting I look around and take attendance. I am not talking about physical attendance, but virtual attendance. Who is really there, and who is just coasting? Then again, I may be the one who is coasting. I am out on my bike, backpack over my shoulder, pedaling down the sandy beaches with a gentle, warm breeze at my back. I look up in the air-conditioned meeting room, far from any window, and feel someone's gaze on me. Guilty, I confess. I wasn't there. Anyway, you knew my bike trip was only a fantasy. The wind is never at your back in real life.

The experience related by Dick Hamming about wartime engineering endeavors is a common one among the older engineers that I know. That was the one time in their lives that they operated full out. Miracles were accomplished: radar, pulse code modulation, information theory, cryptography, proximity fuses, the computer. And, unlike my days of meetings, they loved it, one and all. What a tragedy that the only way to remove all those barriers to creativity and production seems to have been a war.

Years ago I had dinner with Jerry Weisner at a restaurant along the Swedish coast. The fog rolled off the sea, while the smoke curled upwards from Weisner's ubiquitous pipe. Weisner was at that time the president of MIT, but some years earlier he had been President Kennedy's science advisor. There was something I wanted to ask him. He had seen and known all the great during the time of Apollo and Camelot. "What distinguished the great from the talented, but ordinary people he had known?" I asked him. He chewed on his pipe, and looked out at the fog. "Energy," he finally said. "The great ones had a source of energy that kept them constantly alert and involved, whereas the lesser people were only occasionally able to reach those levels." (I have undoubtedly paraphrased his idea somewhat.)

In a technical meeting the other day a speaker drifted momentarily from his subject to describe an experience that he had just had. It seems that he had been participating in a bicycle race when a heavy rainstorm began. The brakes on his bike failed, and he found himself at high speed in heavy traffic in a driving rain, and no brakes. But he described the experience as a kind of revelation. He said that his consciousness and kinesthetic sense had risen to a new level where he had a stark awareness of vivid detail. He felt a sense of unity with his bike and with the totality of the experience. With a twinge of regret, he said that he had survived, and had since resumed his customary subdued life.

The rising to new levels of ability or experience is a common one. All of us know how we feel when we first emerge into the world after an illness. We smell the breeze and see the colors in the flowers, but after a day we revert to ordinary existence and drive home from work on autopilot, unconscious even of the route that our car seems to have chosen on its own. We all know these things, but what we often do not know is the degree to which our professional work can be enhanced by either necessity or unique fertile conditions. The engineers who worked on wartime projects in WWII remember, and they all were changed forever.

Entropy in Modern Life

I have always been fascinated by the concept of entropy, both in the theory of information and in the second law of thermodynamics. Entropy is a measure of disorder in either case. In Shannon's information theory, entropy represents the intrinsic information content of a source. There is no information in regularity, which is totally predictable, whereas an unpredictable source carries a great deal of information. In thermodynamics entropy has to do with atomic disorder. The second law of thermodynamics states that entropy is always increasing. Disorder will eventually reign supreme. Allow two gases to mix, and they will eventually reach a state of equilibrium of maximum entropy.

As a philosophy of life, entropy has an appeal to me. Uncertainty grows, people get busier, and the heat of life is being ever increased. On the micro level atoms are jumping about ever more frantically, while on the macro level so are we. I am drowning in mail, magazines, messages, and the information cacophony of our time. I am ashamed of myself, but I can no longer return all the telephone calls I get, nor answer all my mail. The irony is that at this moment I am awaiting the return of a call that is very important to me, but I have the certain feeling that the memo of my call has been trashed by the person whom I await. Truly, the world doesn't work very well — we are too busy.

The essay "The Frenzied Life" was written for *Science* magazine. It is the only essay in this book that was not originally written for *Spectrum*. It was commissioned by my friend Ellis Rubinstein, who also helped considerably in revising the original manuscript. I hope that you do not relate too much to my frenzied existence, but keep in mind that somewhere out there the heat is being turned up. Sooner or later all of us will feel like the proverbial cat on a hot tin roof.

"What's Going On" is in a similar vein. I cannot shake the feeling that for all my running about, there are things happening that I should know about, but of which I am unaware. Every time I go away on a trip I worry about what is happening back in my office. It is almost the opposite of the old question about whether a tree falling in an empty forest makes a noise. When I am away I hear in my mind the echoes of many falling trees back in my office. Of course, if I am in my office tending to administration, I worry that things are happening out there on the road. Wherever I am, things are happening somewhere else, and I won't know about them.

There are two stacks of papers and magazines piled high on my desk. I put them there because I am sure that there are things happening in those magazines that I should know. Someday I intend to read them. Periodically I throw the piles away and start new ones. Actually, I confess that I have my secretary throw the piles away. I am too afraid of uncovering something vitally important, but overdue by months, in the midst of the stack. Perhaps I might see a letter that says that I may have already won a million dollars, if only I call a certain telephone number by a certain date, now well past.

The mail at home is no easier to handle. I have conceived the idea of solving the nation's energy problems with simple mail slots on the sides of houses, where the letter carrier drops the mail, and it falls directly into the furnace so as to heat the home. As it is today, I have a difficult time closing the loop between the post office and the garbage disposal people. Whenever I trot my garbage cans out to the street, suitably gift wrapped, I invariably see some corner of an envelope in the trash that looks like an important bill or a check. I try to ignore the questionable envelope, but the visual pattern eats its way into my brain. In the end I stop wheeling the garbage cans and pull out the letter. Usually I find that I have won a million dollars again. Occasionally I find that I have only won a large-screen television set, and all I have to do is go pick it up at some new housing development. I know that there is something wrong with this, but I can never figure it out. You want to know where the real quality research is being done in this country? How about the junk mail people?

Maybe a shrink could help me with my paranoia, but while I was lying on his couch I would be missing out on things happening in the outside world. I would have to explain that when I was a youth I had to practice my violin, while other kids were out there playing football and baseball and doing important things like that. Little did I know at the time that this would be a precursor of an information age paranoia. Now the things that are going on have to do with the accumulation of information. Some people are in the know, and some aren't. It is important to be in the former group. Some day soon *People* magazine will have a special issue featuring the 50 most-in-the-know people. There is some danger that my picture will be in the back of the magazine among the 30 most-out-of-the-know people.

There is one time that I feel absolutely safe from information exclusion, and that is the *snow day*. There is no more comforting and glorious concept in modern life than the snow day. People in southern California and other warm climes may not relate to this, but the idea is fairly simple. One morning you wake up and the ground is covered with snow. The phone rings, and someone tells you that they have declared a snow day. Everything for the day is cancelled. Everything. No one is allowed to work. No one is allowed to visit other people and exchange information. No one is allowed to generate, process, or communicate information. Nothing is allowed to happen. The only thing you are permitted to do is to sit in front of a fire and watch the snow flakes drift by the window. It is wonderful.

The problem in recent years has been the pernicious effect of global warming on the frequency of snow days. The physicists are probably doing something that they shouldn't. Anyway, we do not seem to have many snow days in New Jersey any more. So I have a proposal to restore the splendor of this wonderful concept, and to make it more generally available throughout the country.

My idea is that every morning of the year the President, or his delegate, would spin a giant pointer — probably at 6 A.M. on national television during a station break. Maybe Vanna White could assist. Out of 365 sectors on the wheel perhaps two would be specially marked. If one of those special sectors came up, the President would declare a generalized, national snow day. No one would be allowed to work, etc. There would be severe penalties for anyone caught in the act of making things happen. However, you would not necessarily have to sit in front of a fire. You could, for example, go to a beach. Whatever terrible things you had scheduled for that day would be cancelled. This way, no matter how bad a day you had in store, no matter how much you had dreaded a particular day, there would always be hope. It might be a snow day.

Lacking snow days, the complexity of life escalates daily. In addition to the mounting demands on our overtaxed minds posed by the accumulation of information, there are the little things that sneak into our lives, like the need to memorize ever-longer telephone numbers, PIN numbers, passwords, and so forth. Every service provider feels that his service should have its own identification number. When I look back on my essay "The Information Age," I am struck with how much worse things have gotten in the decade since this was written. In that time I have accumulated a garage full of bicycle locks whose combinations I have forgotten. I can't throw them away, because they are good locks. I know there is not much of a market for them, but I cannot bear to throw away a good piece of hardware that only lacks the proper software to be functional.

I have a savings account where a PIN number is required in order to make a transaction. I chose my PIN number very carefully to be related to something that I would never forget. I cannot remember what that something was. My account is now very safe; no transaction has been made for a very long time.

Every time I log on to a computer I cringe, dreading the time that I will receive that hated message, "password has expired." The systems administrators have, in their infinite wisdom, provided help for people like me. A friendly message comes up on the screen telling you how to choose a new password. First, it says, you should be sure that your password is easy to remember. However, it admonishes, you should not use your address, telephone number, social security number, birthday, name, or anything personal that other people could discover. Furthermore — and now the friendly language turns darkly threatening — you should not use characters that form words, the use of vowels is discouraged, and there should be a variety of nonalphabetic characters, like -:#@^. Just be sure that it is easy to remember, though.

This last essay closes with a little dialogue on "cheap bits." This little paragraph was my first venture outside the confines of comfortable technical writing. Many years ago I wrote a paper about the implications of inexpensive broadband communications, and I gave it some title with the term cheap bits. I cannot remember what came into me at the time, but I included this little dialogue as a preface. Of course, all technical papers have to be cleared by my company, and when I submitted this paper for clearance it was denied. The anonymous reviewer said the preface contained material that was insufficiently formal and unprofessional.

No one has ever accused my company, or any other large company, of having much of a sense of humor. This brings to mind one last story. Back at the beginning of my career I was asked to give a talk to a group of visiting firemen. They weren't really firemen, of course — I just thought of all visiting groups in that way. This particular group was the Young Millionaires Club. Naturally, I felt a great empathy for this group.

I remember sitting through the presentation before mine, and feeling the stifling atmosphere in the presentation room when I approached the podium. I was relatively new at the company, and had given very few talks. Millionaires made me feel uneasy. I have since forgotten the context in my speech, but in response to some question about regulation I observed that "the Bell System would be better off if it did away with Bell Labs and used the money to buy an FCC commissioner."

When I made this offhand quip, the public relations person who was hosting the group strode to the front of the room and announced that this talk was being terminated. He turned to me and said loudly, "You will never give another talk to a group of visitors." I said the first thing that came to my mind. I said, "Is that a promise?" As the group filed out they all shook my hand. Millionaires were really all right.

The Geniuses Among Us

Whatever became of all those child geniuses we heard about when we were young? Remember the stories about the kids who could multiply 12-digit numbers in their heads instantly? Or the whizzes who graduated from high school at age 12? The tabloids were filled with such legends, and it seems these young geniuses were often headed into computers or electrical engineering. So where are they now?

Fortunately, when I grew up I found that the world was not cluttered with these dangerous purveyors of brilliance. Meeting the famous engineers I had read about in graduate school, I slowly discovered to my relief that, figuratively speaking, they donned their pants just like all the rest of us — that is, using the conventional head-first technique. There was only one among my associates who retained his lofty genius status. I knew he was a genius because he rode a unicycle and was able to juggle six tennis balls simultaneously — *QED*. One day he sidled up to me and confessed in a whisper that the place scared him because there were so many smart people around. Wildly I looked about me. I didn't see any smart people. There were just my ordinary friends and enemies. It was a moment of great insight. I forget what it meant.

In graduate school I took a course in something I have since forgotten about. The grade was to be determined based solely on a lecture delivered by each student on some paper in the current literature in whatever field it was. One woman in the class delivered a particularly esoteric paper for her assigned lecture. With brilliance and flair she filled the board with equations of deep significance. Finishing breathlessly, she stood back from the panorama on the blackboard and asked if there were any questions. She looked like Picasso in front of *Guernica*. Overcome with a deep sense of inadequacy, I was paralyzed with dumbness. As she headed back to her seat, one person in the class hesitatingly and timidly asked her about the first equation on the board, which in fact was merely a basic assumption. The question revealed a vast ignorance in the questioner. In the ensuing embarrassed silence we stared at our classmate, now vulnerable and alone in his intellectual birthday suit. (I edged my chair away slightly; this might be contagious.) Near hysteria, the lecturer suddenly burst out with "I don't know! I don't understand any of this stuff. I only memorized all those equations!" It was a moment of great truth. I'm not sure what to make of it.

Once I attended a thesis review given at our laboratory by a young Ph.D. candidate from the Massachusetts Institute of Technology. The thesis review is a talk about the candidate's Ph.D. thesis. It is an opportunity for the candidate to present himself as a "can't miss" job prospect, to display a becoming humility, and to screw up. This particular candidate was destined in later life to become a genius. We didn't realize it at the time, but there were telltale signs — he spoke incomprehensible English, and the talk was so poorly organized that nobody knew what the topic was. Among the audience was an eminent mathematician (EM). I had never met this EM, but his fame was known far and wide in the kingdom. I was privileged to be in his presence. After the talk he asked a question. This question was insightful and brilliant; it cut to the very heart of whatever it was. I sat in awe of this level of comprehension, so in contrast to my own primordial state of ignorance. After a moment's hesitation the young candidate strode down the aisle of the conference room and stood before the chair of the EM. Towering over the now-quivering EM the candidate shouted, "If you really understood this, you wouldn't ask a stupid question like that!" I shivered at my near escape. I had learned much. I have since forgotten.

Sometimes I awaken in the predawn — the hour of the wolf — when there is no one in the world but me. I worry. There is no shortage of material to worry about. There is a morrow coming. Soon I will have to get up and don my engineer's disguise. I have perfected my role, but maybe today I will be unmasked. I'm not worried about my friends and associates. I can see the fine line of their wigs and the telltale mascara. But somewhere are all those geniuses. I heard a rumor the other day that they are in Japan. I am worried.

Diminishing Dinosaurs

Many of us know what it is to arrive in Europe after a sleepless overnight flight. Recently I found that, as usual, my hotel room in London would not be ready until later in the day. Unshaven and groggy, I sought refuge in the nearby Natural History Museum. I collapsed on the nearest bench and found my perspective dominated by the skeleton of a large dinosaur. Large, yes, but nothing like the size of the dinosaur I used to visit in the Carnegie Museum in Pittsburgh when I was a youngster. Now *there* was a real, giant dinosaur. Was this the best a world capital like London could come up with?

After a long while the museum guards started to eye me with increased interest. Time to move on. My curiosity was barely sufficient to cause me to detour so as to read the small plaque at the base of the dinosaur. It said that this skeleton was only a plaster of paris cast, but nevertheless a *full-size* replica of the very dinosaur I had known in Pittsburgh. I stood there for a long time, thinking of what it is to grow up and find the world not as big nor people as omniscient as they appear when we are children. We spend a lifetime trying to outgrow those images. There was no giant dinosaur after all.

People are not really so big or so smart either, and though we think we know our own field, we often believe that giants surely exist in other fields. I was at a meeting of some scientific committee in Washington, D.C., where we were debating national policy in computer-related research. It was very easy to become bloated with self-importance. I could feel the level of pretentiousness rising around the long, wooden table.

Soon, my eyes rose to the oil painting on the wall, representing a famous American scientist-statesman doing something wonderful. Possibly he was decreeing that henceforth there should be electricity. Maybe our committee could do something similar, and become the subject of a similar painting. I was thinking of how I could adjust my schedule in order to sit for such a painting, when my train of thought was broken by the attention-getting remarks of a famous professor. "What are we doing?" he interjected. "Don't we realize that we are only a bunch of amateurs at matters of national policy?"

The word "amateurs" was an indictment that struck home. A pall descended over the conference table, like an ominous February snowstorm. Coherent thought was blanketed with remorse. How could we dare? Had we forgotten our place? It took all of my energy to shake off momentarily the induced lethargy, but I had to voice it. "Who do you think is running the country?" I asked.

Now I imagined a new meaning in the stillness that blanketed the table. Deep contemplation, perhaps. But no one responded. Yes, we were amateurs, but who else was there? My gaze returned to the statesman in the painting. Certainly he was a giant of my youth, but as I recall he was also an amateur at starting new countries. He did all right anyway.

From this perspective I can have sympathy for the IEEE's U.S. Activities Board and similar organizations whose task it is to advise governments. We all may be amateurs, but so are many of the designated leaders — particularly when it comes to matters of technology. We should not be ashamed, for there are also amateur leaders of armies, and even (I whisper this) amateur managers of corporations. I remember the first time I ever had the privilege of interacting with some top managers of a large corporation. When I returned from my visit to that Valhalla, someone asked me what the great experience had been like. Without thinking, I responded that it had been like being on an ocean cruise on a great liner, and being asked to the bridge to meet the officers. But when I reached the bridge, I found no one there. The noises from the parties of passengers below drifted up to the empty control room, and the great ship sailed inexorably on. I had really expected to find giants.

The prevalence of amateurs is a comforting thought to keep in mind. Unfortunately, it is often true that we ourselves turn out to be amateur electrical engineers. Not realizing that possibility, we seldom welcome suggestions from outsiders about our profession. Yet how often those inexpert opinions and ideas turn out to be right. It is an embarrassment that most experts try to avoid by making a preemptive strike. "It is impossible," the expert pronounces with finality. A lesser expert is suppressed, but the amateur plows ahead. It is a good thing there is a plentiful supply of amateurs.

There is a fine line between being an expert and being an amateur, a line easily crossed at any time. In technology we do not even need the Peter Principle. It is unnecessary to be promoted to get to the level of incompetence. You merely sit and wait a little while, and it will come naturally as technology changes. You may then be irredeemably incompetent, or you may just become a member of that relatively harmless group whose fields of expertise have faded from importance.

With so many of us losing our expertise or suffering the paling of glamour in our fields, it is fortunate that there is a market for used experts. There is nothing wrong with being an amateur so long as you realize your status. There is a great need in the world for generalists. This need is particularly acute in engineering, where the splintering of our profession into so many complex specializations has left a patchwork quilt of nonintersecting branches of knowledge. Someone has to walk those cracks, and that someone must be a generalist and, of necessity, an amateur. There are no expert generalists.

Much as we need generalists, the only path to being a successful generalist is through being an expert specialist. People will listen attentively to amateurs in any field so long as they know the amateur is, or has been, an expert in something else. I do not know why, but that is the way the world is. Get credibility somewhere and it converts everywhere. But here I am, writing as if I knew what I'm writing about. You should pay no attention. I am, alas, only an amateur columnist.

The Dreaded Test

Last night I had the dream again. I was back in a surreal college, and it was the day of the final exam in an important course. The only problem was: I had never attended a single class in that course. It is always a little unclear, as dreams can be, whether my failure to attend the classes was because I could not find where they were held, or because I had forgotten I was enrolled in that course, or what. A sense of foreboding hangs over the dream as the dreaded test looms near, and I am engulfed in helplessness. My relief is palpable when I wake and realize that all is well. The last test was taken many years ago, and now my degrees are safely hidden away. I will not give them back.

I used to think I was the only one with that recurrent dream, but apparently it is very common. I have even perfected a variation in which, although presumably I have attended the classes, I am unable to find the location where the exam will be held. I wander from building to building as the time for the exam comes and passes. I am sure there is some psychological interpretation, but I do not think I want to hear it. All of us know the apprehension engendered by an approaching exam. The curious part is how that narrow slice of time we spend in college is forever fixed in our minds, and people like me are forced to replay that apprehension the rest of our lives.

When I finished my thesis, the last hurdle of college was completed. On a job interview I mentioned casually to one of the company representatives how relieved I was to be able to put the thesis work behind me. I will never forget the funny way in which the man looked at me and said, "Do you realize that the rest of your life will be an endless succession of theses?" It was a moment of truth. No, I had never looked at it that way. I thought I had completed my last thesis, taken my last course, learned all the technology I would ever need, and endured my final, final exam. Alas, the ugly realization belatedly dawned — there is no last test.

Life is full of tests. Every day we go out into the world faced with the necessity of learning as fast as we can, and of suffering tests on that knowledge. The differences are that the tests are not multiple-choice, the grades do not come so frequently or so obviously, and there is no textbook that contains all the answers. In my dreams there is never an implication that the tests are too difficult, or unreasonably constructed; the fault is always my own. However, in real life there is no guarantee that all the tests will be fair — only that they will exist and that we will be treated according to the outcomes.

In thinking about tests I always take heart from an incident in my childhood. I took violin lessons from an old, poor Hungarian immigrant. I always imagined that he had been a great violinist who had fallen on hard times. Whether or not he had ever been the great virtuoso of my imagination I will never know, but it was evident that he was poor. He lived in a tiny apartment in a low-rent district, and always painstakingly counted out in advance the change he would require for the bus he took to our house to give me my lessons. As children can be, I was callous about his plight, and only mildly surprised that not everyone lived in a house like ours.

The family that lived across from us had become suddenly wealthy for reasons that I have forgotten. They had two small daughters that they now desired to be educated musically, and they enquired of my mother if my violin teacher would take them. Before my next lesson, my mother confronted the old teacher with the two little girls, saying that they wished to take lessons from him also. However, the teacher said that he would have to give them a test before he could accept them. They would have to demonstrate musical ability.

The teacher plunked a note on the piano. "Hum that note," he asked one of the girls. He repeated the test with the other sister. Then he turned to my mother. "I cannot accept them," he said, "they have no ability."

My mother said not a word. She took the old gentleman by the arm and dragged him into our kitchen. We heard hushed voices for a few minutes. They returned, and he looked at the two frightened little girls. "I accept," he said.

Many years later, both of those little girls made their living playing with major symphony orchestras. Needless to say, I never did. I have often wondered about that test he gave them.

Back to my own story — the very next interview trip I went on, after hearing how the rest of life would be spent writing theses without end, illustrated the nature of this purgatory. I was visiting the research lab of a large corporation. The first stage of the day of interviews was to be a talk by myself on my thesis work. It was, of course, a test. However, the supervisor of the group that worked in my speciality was going to be unable to attend my thesis review, so he had scheduled an early meeting with me before my talk. "Tell me about your thesis," he asked. I went to his blackboard and told him about what I had done. I was nervous about the upcoming talk, but not about this informal discussion.

After I finished describing my thesis work, the supervisor looked at me sternly. "I don't think you can make it around here," he said. They cancelled the rest of my interviews. I had failed, and I had not even known it was a test.

Sometimes I think about that interview. Perhaps the supervisor was right. On the other hand, maybe whatever criterion he applied was the technological equivalent of "Hum that note." Many years later I ran into this same supervisor, now a high-level executive. "Haven't I seen you somewhere before?" he asked me.

"No," I replied.

There is no end to learning, and no end to tests — so my dream will continue. Sometimes even in real life we have final exams in subjects where we have to forgotten to attend the course. Best of luck, then, in whatever dreaded test you face today. Just hum your best note.

The Footsteps of Giants

Every now and then I sit down at my piano and try to compose a tune. I start out hopefully, but no sooner have I played "plink, plank, plunk," than I recognize once again that I am a failure as a composer. In the natural process of trying to rationalize my ineptitude I lash out mentally at Bach. He had it easy, I think. Any tune he picked out at his harpsichord was original; there just wasn't that much around in 1700. I can just imagine his mother listening from the kitchen. "Little Johann has done it again," she says contentedly to herself as she hears his latest "plink, plank, plunk." Oh, I am envious!

Perhaps it is the same in our profession. I have a rare copy of the 1952 Bell Labs directory. (In electrical engineering that is ancient. The directory itself is scribed on a rolled up papyrus.) What is astounding to me about this directory is that it seems that almost *everyone* in it became famous. You may well find something similar in your organization too, if you are able to trace it back to this prehistoric time. I find this incredible. Perhaps the world was a much simpler place then, where every "plink, plank" of technology resulted in a great invention. Or were our predecessors giants, whom we have no hope of matching?

On only the first three pages of my old directory I see a profusion of legendary names; I see four people who went on to win Nobel prizes, four who would become presidents of large corporations, a college president, a dean or two, and numerous vice presidents. I see the inventors of the transistor, the laser, information theory, coding, and negative feedback. What kind of a world was this, where famous names were disguised at the time as nobodies? Did these people realize they were on their way to fame and fortune, or were they just like we are today, lost in the feeling of muddling their way in obscurity towards the unknown?

In contrast, when I look at the 1986 Bell Labs directory, the names on the pages swim before my eyes. I am reminded of the immortal question of Butch Cassidy, who kept asking, "Who are these guys?" (Lots of whom are now women, by the way, so score one for the new generation.) But here is the question: If I were to look at this same 1986 directory in the year 2020 (another 34-year time lapse), how would I see those names? Would most have become famous? Or would they just be so many forgotten names in an old yellowed booklet, like pictures in a dusty yearbook from somebody else's high school class?

My own generation has had two decades to prove itself. Many of my contemporaries have made names for themselves, but somehow they don't seem as notable as those illustrious names of the past.

I took over the chairmanship of a volunteer board recently, where an oil painting of one of my predecessors dominated the room in which we met. Theodore von Karman looked over my shoulder, and I knew that John von Neumann used to sit there, at my right. Don't expect me to live up to them, I thought. None of us is up to it. Yet how much harder it must seem for the new generation: tens of thousands of new EEs starting their careers in a world laden with complexity and filled with large projects that subsume individuality. Is it possible for them to be as famous as those names in the old directory?

I can posit two theories. The first theory says yes to the question. If the new EEs wait their turn in line, the inventions, the promotions, the great discoveries, and the fame will follow. We live in an expanding universe where the past always looks like a small group of famous people when viewed from sufficiently far in the future. Technology is infinitely expandable, and as great as the inventions of the transistor and the laser were, there are even greater discoveries awaiting the next generation.

It is in retrospect that Bach is a genius. It isn't really as if every tune he played was an original. What he had to do was break the bounds of a system of musical thought that by our standards was greatly constrained. The composers who followed him were for the most part unappreciated by their contemporaries. Decades from now the modern composers will seem just as great.

My second theory says no. The world has changed fundamentally. Complexity has increased so much that large teams of people are now required to create and design — or even just understand — ordinary systems. Under these circumstances, how is individuality to manifest itself?

Even in music, there is learned argument about whether or not contemporary composers such as John Cage, George Crumb, György Ligeti and their associates will be remembered by posterity. One critic has maintained that their music is now so esoteric that the mean time to acquire the ability to appreciate it exceeds the average lifetime of the listener. Something the same may be happening in technology. "Plink, plank, plunk" just will not do anymore.

Stephen Jay Gould of Harvard recently wrote in *Discover* a thought-provoking essay on the disappearance of the .400 hitter in major league baseball. Even though I am a baseball fan, that loss isn't something I worried about before. Now I do. It seems that batting above .400 for a full season was not all that uncommon in the past. Eight players exceeded .410 in the fifty years before Ted Williams batted .406 in 1941. No one has batted over .400 since. Why have these super hitters disappeared, Gould asks, in view of the well-established fact that athletes have improved in all quantifiable aspects of sports?

Baseball is a delicately balanced duel between offense and defense. In fact, the balance has been so equal that the mean batting average throughout the entire century has been almost constant at .260. As the pitchers have improved, so have the batters, and at a standoff rate. What has happened, however, is that the *standard deviation* has diminished steadily through all these years. In other words, fewer people stand out as exceptional. Theodore von Karman was a .400 hitter. There aren't so many anymore.

Gould offers an explanation for the disappearance of the .400 hitter. Humanity is definitely improving in skills of all kinds, but there is a limit on human performance that moves more slowly. This limit is like a wall to the right on a graph of the distribution of batters or engineers. As we improve, more and more of us are scrunched closer to that limiting wall. The standard deviation narrows, and fewer of us stand out. In other words we too are giants. But we live in a land of Gullivers with nary a Lilliputian in sight.

If you sometimes despair that the landscape is too cluttered with the footprints of giants, remember that your footprints are even bigger. It is just that there are all these other big shoes today, everywhere we look.

Faking It

One Christmas I got this giant book of all known piano music. It is, alas, a "fake book." Only 50 bits of information accompany each song reproduced. The ultimate in audio compression has been achieved. It is only necessary to run these 50 bits through a suitable processor (a talented musician), and the original music is reconstructed in its entirety, with nary a missing subtlety. Unfortunately, the algorithm for reconstruction is not contained in the fake book. Sometimes life is like this. I have only 50 bits to work with, and a great deal is being required of me. I have developed over the years a primitive algorithm designed principally to get by under these circumstances.

I don't remember where I heard this story, and it's probably apocryphal anyway, but it seems that an art dealer in Paris came across what he suspected was a work by Picasso. In order to authenticate the piece, he drove to the south of France and confronted Picasso with the painting. "It's a fake," Picasso announced. Picasso's agent, who was standing nearby, interjected, "But maestro, I saw you painting this work myself." Unperturbed, Picasso responded, "So, I often paint fakes." Alas, life is like this too. Sometimes I engineer fakes. Sometimes I am required to be, as it were, a fake.

Lest you, the reader, should feel above all this, let me point out that it isn't all that hard to be a fake in our profession. Some years ago an associate and I elaborated on a particular scenario. Suppose you are involved in an accident resulting in brain damage. The doctors tell you that you have suffered a loss of, say, 50 points in your IQ. Electrical engineering is the only profession you know, and you have grown extremely fond of your regular pay check. Even though you're not by any stretch of the imagination (to coin an appropriate phrase) a genius, you still have a certain amount of guile. The question is — how long can you survive in our milieu? The answer we agreed upon? Indefinitely.

First you must sprinkle your conversation liberally with buzz words. The casual mention of such things as microprocessors, VLSI, operating systems, high-level languages, artificial intelligence, knowledge-based systems, CMOS, NMOS, CAD/CAM, injection lasers, RAM, ROM, and so on in any cryptic context will suffice to buy time. Occasionally you should throw in an acronym like "SBDR." The first list of buzz words should be memorized. On the other hand, the acronyms may be generated at random in real time (as I did with SBDR). Half of the audience will recognize the acronym, since it's bound to stand for something. The other half will tune out everything else you're saying, frantically trying to fit words to the acronym. Both halves will recognize and appreciate your know-how.

The next bit of cunning you should develop is the art of asking deep and meaningful questions when you haven't the foggiest notion of what is being discussed. Here a small set of generic questions will suffice. For example, if a logic diagram is being examined, you muse, almost under your breath, about redundant gates in, say, the lower left-hand region of the diagram. If computer code is being discussed, you offer that you're not sure that this program will work under all sets of input conditions. This must be said in such a way that the intimation is that you have discovered at least one such suspect input condition yourself. If you are looking at a circuit board, point to some portion (or possibly the entire board) and ask if this could not be made into a custom chip. If system specifications are at issue, give the opinion that the set of specifications is not complete and may in fact be contradictory. If you are looking at some intended product, shake your head and say softly that the competition can make this at a lower price.

On occasion, you may be invited to higher-level briefings. Here you need more crass cunning. Say you are shown some computer architecture where you're not even sure if the slide is upside down or not. Rubbing your chin, you ask if the designers have considered a distributed-processing approach to this task. Possibly you offer that the design might well have taken greater advantage of the inherent parallelism in the required functionality. When confronted with a discussion on, say, some software system for data processing, ask knowingly what operating system this will run under. On being told, disregard the answer, clearing the few registers remaining in your deficient mentality, and ask if they have considered any alternatives. "Is this the best system you could come up with?" you ask with obvious doubt. If this is said in the correct manner, it should serve to amplify the doubts lurking beneath the calm exterior of any good engineer.

There is one sizable problem. The person you are talking to may also be faking it. The conversation can degenerate into strange and uncharted pathways. The strain on your underpowered mind can be severe. "I notice that this design uses a Z-80 microprocessor," you begin. (This is usually safe, since there will be some microprocessor in there, and you will only be corrected as to the type number.) The other engineer shakes his head hesitantly and replies, "Yes, that is the best operating system for this particular application." Somehow that doesn't seem to ring just right in your primitive memory system, but the ensuing silence forces you to forge ahead. "Have you considered putting this operating system into custom VLSI?" you ask. Now you are beginning to get out of your necessarily shallow depth as the response is "Yes, but the TTL implementation of the operating system contains software bugs." Reflexively, you shoot back with, "That wasn't the way it was on the Snork project." (This project will be classified "Secret," if there are any questions about it.) Not to be outdone your antagonist comes back with, "Yes, but in the Snork project the software was written in NMOS and used excessive power." By this time both parties have reached a high state of tension. Disengagement will come quickly, and nothing will be lost unless an engineer fully competent and conversant in the subject is listening. Then you really have troubles.

Of course the scenario of losing 50 IQ points seems fanciful. But maybe you already have through attrition. Have you had your IQ checked lately? I have this theory that deep inside all of us is this real intelligence. But it's often well hidden. There are layers and layers of me, just like there are in operating systems. Sometimes when you think you're talking to me you're just dealing with an outer layer. I may seem a little slow. Maybe I'm faking it. But even Picasso didn't run full out all the time. Sometimes I'm just running at half throttle. That way I get more miles per gallon. I think.

The Frenzied Life

One morning, heading to my office, I overheard a fragment of conversation between two strangers standing in front of me in the elevator. They were talking about some young person who had just been promoted to something or other. "They say he's the new Bob Lucky," one of them was saying.

It's odd enough hearing your own name spoken by strangers — your ears redirect themselves like an adaptive array — but just combine that with the disquieting experience of discovering you've been relegated to the scrapheap of legends in your own lifetime ....

By now the doors were open at my floor, and as I pushed past the pair, I turned back. "Once I was the new Bob Lucky," I said. The doors closed on two surprised and puzzled expressions.

As I headed down the corridor, I couldn't shake the feeling of being superceded and useless. I was thinking: I wasn't always just another graying manager. Once I had worked at trying to be the new Claude Shannon. Those were the days when I couldn't wait to get up in the morning and get into the laboratory, when my greatest thrill was a new equation or a confirming trace on an oscilloscope. Money seemed to descend from science heaven without human intervention, and the chief management strategy was that of benign neglect. My phone seldom rang and my mail box stayed empty. After all, no one knew my name, much less was described as being the new me.

Reprinted from *Science*, vol. 252, p. 1131, May 24, 1991.

I remembered what a pleasure it had been to work in the lab into the wee hours, but how the weekends were gloriously free to be spent with the family and the lawn. It was a time when people did real science, I told myself — and science for the sake of science, I added pompously. We didn't spend all of our time frantically dashing hither and yon playing diplomat and fighting the frenetic scientific bureaucracy. The thought of how much things had changed threw me into a pool of nostalgia, sure of one thing: this new Bob Lucky was a beleaguered, stressed-out creature who would have been scorned and pitied by his former namesake.

I rounded the corner and spotted my secretary's cheerful face, but my thoughts were still back with that pair on the elevator. I wondered if they would be as bored with me as I was with the old-timers of a couple of decades ago. They used to drive me mad with their endless talk of how things had been in their generation. Those were the days, they'd said of the years during the second world war. No bureaucracy, no budget squeeze, a great cause. They would reminisce about developing atomic energy and the digital computer, about perfecting radar, gun control, and proximity fuses, about conceiving information theory, pulse code modulation, and cryptographic techniques ... and then, they would sigh and lament how when the war ended those budgets came back, and the managers started to manage, and all that good science and great satisfaction went away. How tiresome all these stories were for me at the time, I thought.

But then the Old Bob Lucky that I had become tried to defend himself: the environment for young scientists today is simply not the same as it was a generation ago. There's a greater entropy in the air today, I thought defensively. It's as if somewhere the heat of life were being turned steadily higher. Perhaps we are obeying the second law of thermodynamics: we spin ever faster into a stream of chaos filled with airplanes, paperwork, fax machines, computers, committees, fruitless proposals, and burgeoning information.

Yes, I thought, science is not what it used to be, and an image came to me from a short fiction article I had read almost 25 years before in a popular American magazine. The main character was a scientist who seemed to be on the verge of a great discovery — if only he had the time to finish his work. But suddenly he found himself on dozens of boards and committees. Rushing sleeplessly from airplane to airplane, he kept running into the same set of scientists, who themselves seemed always on the verge of great discoveries, if only they could stop rushing from plane to plane. Now the protagonist — scientist that he was — decided this was a phenomenon that had to be studied. There seemed to be a pattern about it. Perhaps he could regain control of his life by investigating the sources of all these prestigious invitations he kept receiving.

I forget the details of his investigation, but in the end it turned out that the whole fabric of the scientific committee structure was being controlled secretly by aliens. They were a resourceful race: they had figured out that they could keep Earth a relatively backward place by making sure that, whenever someone threatened to make a scientific breakthrough, he or she would be invited to join a blue ribbon panel.

That story was written long ago, but I was convinced, as I put my briefcase down on my desk and frantically scanned the day's schedule of travel arrangements and meetings my secretary prepares for me each morning, that it is more true today than ever. The printed words on my schedule blurred together, just as all the meetings it represented seemed to do. I had my recurrent feeling of being only a package that gets delivered wherever the instructions dictate. I think of it as the "if it's Tuesday it must be Stanford" syndrome.

You've probably read the experts' explanations of this new busyness — the globalization of research capability, the rise of interdisciplinary work, the increased directivity of the funding agencies, the dramatic increase in the cost of scientific research. And everywhere the footprint of the computer. Like some new race of Bigfoot trampling over traditional experimental technique, creating a new methodology of research, and worse — escalating the complexity of our work lives.

I thought of how tiring it was to be swimming constantly in a swelling tide of conferences, meetings, and publications. Whatever happened to the old friendly telephone call? Gloomily, I considered how the abundance of answering machines, computer mail, cellular phones, fax, Federal Express, and bulletin boards had rendered that friendly old telephone network obsolete. The trouble is there's hardly anyone home anymore. Everyone is on an airplane. Indeed, the mean position of scientists may be rising steadily — not in social position, but in altitude.

I cast an uneasy glance at my own telephone, poised like an alarm clock at daybreak. It had already been busy this morning with a stack of voice mail messages, and I could see over on my terminal that there were 17 unread messages in my E-mail, all of which got me to ruminating over all the scientists swarming about in my life. Once upon a time, you didn't have to know more than, say, five: a pair of eager post-docs, your lab assistant, your boss, and your principal competitor. Now there were thousands whose names I was supposed to know and who would try to contact me every week or who would pop up at the end of one of my talks and act like we'd E-mailed for years. I love it when the academicians start bemoaning the coming manpower shortage in science. They keep telling us that there has been a drop in high school students electing a scientific career, but if there is, I'd attribute it to the same reason Yogi Berra gave for the unpopularity of a certain restaurant: "Nobody goes there anymore; it's too crowded."

These days there are hordes of scientists engaged in a frenzied quest to beat each other out for funding. In my misery, the image flashed before me of pigs at a trough. All the hogs are lined up, but the trough just isn't long enough. Everyone jostles about, and the little piglets are jammed out.

And just about the time we all seem to have our proper places, a giant rhinoceros carrying a banner that says "big science" or "super something-or-other" sucks up half the feed in the trough.

My secretary was at the door with an apprehensive expression. For an instant I had the scare that I was supposed to be somewhere else right then. I hadn't missed a meeting yet, had I? But she was waving one of those illegible faxes that run my life — an overdue request (read "demand") from somewhere mystically high in the bureaucracy for yet another inventory of research expenditures. This inventory would undoubtedly be compared unfavorably against the latest inventory of research results.

Where does this constant stream of demands for accounting come from, I wondered. Is it the aliens again? But who needs aliens when we have our own dreaded bean counters? Surely the bean counters constitute one of the truly malevolent aspects of the change in our lives in science. Their incredible, inexplicable rise to power! In my overworked imagination I could just picture the neon marquee announcing science's latest horror film: *The Invasion of the Bean Counters.*

In the opening frames of my movie we see MBA-toting, red-suspendered professors piling out of their latest-model 1970s Oldsmobiles, hustling to their classrooms where they are turning bean counting into something they call a science. It looks innocent, but that is the hallmark of the best of the horror genre. Before long these accounting professors have set loose upon the unsuspecting world a sect of apostles whose only religion lies in numbers. By the time the movie draws to its inevitable conclusion, vision and leadership have been replaced by the ever more exact accounting of beans. The viewers leave the darkened theatre for the streets outside telling themselves that it's all right — it was only a movie.

Eventually, I find, even the worst depressions pass. After all, even the most creative minds eventually run dry of apocalyptic visions. At this point, one's mind often ALT-F10s to another screen where one's favorite escape fantasies reside. It was over coffee at around 10 a.m. that morning that my mind turned to devising the perfect job in this seething world of big league science. This old Bob Lucky, I thought, would stun Boss Arno by messaging him that he was opting out; that he was leaving to take a papal chair at the far away and little known University of Shangri-La. The immensity of the grant they'd promised me, I would explain, would leave me completely free to do whatever research I wished to do. The vision carried an extraordinary piquancy — like when Tom Sawyer secretly came back to watch his own funeral. There I would be at U. S-L, but able magically to see my friends back in the Labs speaking of me in hushed tones, voices cracking with emotion, eyes misting over: do you think we'll ever hear from Old Bob again? I am the stuff of legends, I muse; no one comes back from the great scientific beyond to tell us what lies outside our own frenetic sphere.

Sated with this self-pity, I finally turned outward: could the scientific community be spiraling inevitably toward a heat death? If so, what might be done about it? That was when I came upon a modest proposal. Someone once suggested to me that each graduating scientist be given ten chits constituting permissions to author papers. When these were used up, you would have no choice but to join the mute ranks of those who only stand and wait. Obviously you would use your chits carefully; very few minor papers would be published. At one stroke we would condense and upgrade the literature. Information pollution would be contained. We would have time for the spouse and kids again.

Suddenly, I realized the true genius of this system. You'd apply the same technique to membership on committees, invited talks, proposals, etc. Upon graduation you would get a whole package of multicolored chits like a Disney World admission booklet. This would give new meaning to the concept of career planning. For example, you are invited to join the visiting committee at Podunk University: suddenly you hesitate. Not that you mention this out loud, but what if you were chitless when the National Research Board calls?

I was undecided as to whether there should be an open market for chits. Perhaps you should be allowed to negotiate trades. For example, out of desperation you might be forced to trade two invited talks and three committees for a proposal chit. *Science* could list bid and asked quotes for various permissions, I thought. There might even be a market for futures. The free market forces would dictate exchange rates across nations and disciplines, and somewhere there would be a worldwide Chit Board (blue chit required for membership, of course) that regulated the printing and commerce of scientific chits.

The chit system would give us control of the environmental temperature. The more chits issued, the greater the entropy in our profession. Occasionally the Chit Board would make headlines by deciding to devalue chits or whatever. Unfortunately this uncertainty in itself would be a source of added entropy that would have to be factored into decisions.

The chit system may sound extreme, but that day, after my trauma in the elevator and considering the forces that seemed allied against me, I figured that the most draconian measures were in order. And today, thinking about it in this respite between meetings that we call an airplane, while I reformat my overdue report as the stewardess passes me a cardboard dinner, I still think there's some merit to this idea. But maybe you have a better proposal. After all, anything would be better than having our lives controlled by those aliens that run things now. By the way, a piece of advice: I'd be sure to examine your next prestigious invitation very closely. I just received one to speak at a Meet the Faculty dinner at Shangri-La U. It was signed "The New Bob Lucky."

What's Going On?

I picture in my mind the daily travails that once filled man's workday — the hunt for shaggy, long-toothed animals, the tending of thirsty, life-supporting crops, and the sweaty manipulation of clamorous machines. But all that has passed. Today, in the midst of the information age, man fills his workday with the unending and hopeless task of trying to discover what is going on.

Just as in ancient times, my family supports me in this difficult and dangerous quest for bread to put on the table. As I go out in the morning, my dog makes a frantic rush at the front door to try to follow me. It is in part the last vestige of his hunting instinct, but he, too, has the sense of this information age imperative. He knows that things are happening out there in the great beyond — things that he will miss out on by staying at home. I can see it on his face. But no, I must face the uncertainties alone. Amidst all the dangers of the modern world, I must piece together what is going on.

When I return at night, the first thing my wife asks is, "What's going on at the office?" I usually reply, "Not much," but that belies the train of reflex thoughts that her question has provoked in my numbed brain. When my son breezes through, his first and only words are, "Hey Pop, what's up?" Obviously, they are both deeply concerned.

After all, what *is* going on at the office? Another day has passed in meetings, telephone calls, fax messages, electronic mail, tons of paper mail, magazines, journals, and chance meetings with associates. Something is going on — for sure. But what is it? A wisp of suspicion in my mind threatens to congeal into the ultimate disgrace of the information age: I may not be in the know.

Vacation times are especially perilous. Vacation is an unnatural state in the information age. While information continues its inexorable exponentiation — doubling every eight years or so — I remain temporarily stagnant and incommunicado. At night during these respites, I have the recurrent nightmare that I may never catch up once I get back. Some of my friends are already beginning to take fax machines on their vacations. I see the worry on their faces, too. They may miss out on something.

Even driving to work is a time of vulnerability. I used to listen to music and book tapes on my car stereo. When I mentioned having listened to a particularly interesting book to an associate, he looked at me with skepticism. "How can you listen to tapes?" he asked. "What if a nuclear war started, and you were out of touch because you were listening to recorded tapes?" Ever since he pointed this out to me, I have been uneasy about playing tapes in my car. I mean, you never know; something drastic *could* be happening.

Now I notice that my friends are installing cellular telephones in their cars. I see them pass me as they talk on these carphones. There is always a knowing look on their faces, and I am excluded. Probably something is going on.

As hard as we scramble to keep abreast of things, this is still the easier part of our job — the data collection. The data processing is much harder. The things that we observe are the things that are *happening*.

The question is: why are these things happening? For example, there is an organization change at work. The announcement goes up on the bulletin board. Then the speculation begins. What is the meaning of this change? Who and what are in? Who is out? What is the subtle message being conveyed by the bulletin? What is going on?

The management is frustrated by these queries. People should get back to work, they think. A second notice goes up. "The following is why this organizational change is being made ...," it says. However, the whole scenario is infinitely recursive, for no sooner does the second notice go up than another round of questions begins. "Why do you suppose they are telling us this?" people ask. What is *really* going on?

All this frantic activity demands a certain robust level of going-on-ness, as it were. In truth, at any given moment, there may actually be nothing going on in a particular organization.

This, however, is an unacceptable situation; in much the same way nature abhors a vacuum, these days the business world abhors an information vacuum. During this period when an insufficient number of things are going on the speculation rises to an abnormally high level, perhaps in compensation. "What do they mean by this lull?" people ask. "Nothing has gone up on the bulletin board for the last two days," someone worries aloud. "This is surely a sign of major forthcoming revelations."

In addition to worrying about the nature of the workplace environment, the overburdened engineer has to be concerned with what is going on in technology. This is especially difficult, because most of his fellow engineers try very hard to conceal the necessary information. They publish papers in the various journals that deviously lay false trails for the readers, trying to persuade them that their papers are really part of what is going on, when in fact they have nothing to do with anything of importance.

But how are we to deduce this? The trade journals contribute to the general entropy by publishing interviews with managers and famous engineers, who give their opinions about what is going on in technology. Unfortunately, they have no idea themselves, so they make up things.

Somewhere out there, things are going on. Inventions are being made, concepts are being formulated, theorems are being proved, methodologies and processes are being conceived, markets are rising and falling, and styles are being shaped. But these things take place in a cacophony of false starts. Technology is like an orchestra tuning up. If you listen hard, you can hear little fragments of recognizable pieces. But if your attention lapses, you hear nothing but noise.

If you find out what is going on, please let me know. I am desperate.

The Information Age

The advertisers assure us that we are about to plunge into the information age. Exactly what this is I'm not sure, but they say that the era to come will be much better than the Ice Age. Information will be good for us, they add — which is fortuitous since we don't seem to have a lot of choice in the matter. Like it or not we are about to be innundated by streams of information-carrying bits. Information access will be available everywhere — in our homes and offices, through public terminals, and even in our cars and on airplanes. Everyone — excepting those unfortunate individuals and nations who lack the necessary clout in computer IQ — will have the right to bathe daily in tubs of bits, and afterward to pull the plug and watch the excess information drain away. Our time will be spent either accessing information or pretending to access information — the latter activity being recommended to preserve the illusion of the savvy information-age citizen. We shall have to be diligent about this daily hunt for information, since it, rather than material goods, will be the basic commodity of the land. Our position in society will be determined solely by our market position in bits.

I'm a little nervous about the coming of this information age for several reasons. First, no one has told me what I'm supposed to do with all those bits. Moreover, I secretly wonder who will be generating all those bits that the rest of us are forced to absorb.

I barely have learned how to take care of money. What happens if we go on the bit standard? Will I be able to find ways to invest my hard-earned bits? Or can I hide bits in my mattress, against the rainy day when I find myself bitless? How will I ensure that I have enough bits to carry myself into retirement, when I no longer have the strength and ability to conduct the arduous, daily accessing of information which will be the sine qua non of the coming age? Will bits depreciate because of bit inflation, or like fine wines will they improve if left in a dank basement? (Perhaps no packet should be opened before its time.)

What scares me most about the coming of the information age is the implication that it is on its way and not actually here yet. Frankly, I'm having some difficulty handling all the information that flows into my office now. I did a survey not so long ago and determined that my daily office mail contains, on the average, some 300 pages of material, not including all the technical journals I'm supposed to read to stay competent in my profession. That represents about a half a megabyte of whatever-it-is. Far from needing this oncoming information age, I wonder if we should find out who is responsible for generating the current glut and put a stop to the excess information flow before it really gets out of hand.

After Claude Shannon and others conceived the principles of information theory in the late 1940s, a number of studies were conducted to determine the channel capacity of a human being. These studies were fairly consistent and depressing. No one does them anymore. It seems that human beings are only good for about 150 bits per second of input/output. That is all the information we can take in or put out. I ask you, is this a creature equipped for the onslaught of the information age?

At my modest 150 bits per second of capacity, those 300 daily pages would take me 8 hours to absorb, should I be so foolish as to spend my entire working day operating in an input mode. But the real problems begin after the completion of the input cycle, assuming it has been completed. What has happened to those bits? Are they stored away somewhere in my brain for future access? Have they been processed and used to generate new, useful, salable data, or have they just overflowed my inadequate input buffer facilities?

I get a kick out of all the ads in the newspapers and popular press for personal computers that are so technical that people who haven't the faintest idea what they are talking about run around bragging about how many "K" of memory their system has. Well, maybe the same measure applies to people's memories, too. For all I know, I'm working under the handicap of having only a 48K capacity upstairs, whereas you may have a deluxe 256K. So when I blithely shove that daily half megabyte into my economy-model brain, something is guaranteed to fall out the other side. This is the well-known pigeon-hole theory of the brain. Mine may have been used up long ago merely under the trickle of bits preceding this oncoming information age.

The other day I stopped to make a phone call from O'Hare International Airport in Chicago to my son in college. It was about time, I thought, to learn his number by heart, so I memorized it on the spot. Then, after dialing his number, I heard a friendly prerecorded message ask me to dial my credit card number. I confidently put my finger forward to input the necessary 14 digits, long since burned into the read-only-memory section of my mind. My finger wavered futilely while my whole system looped. My credit card number had been zapped from my mind. It must have fallen right out the back while I was shoving the new digits into the front. The inadequacy of my ROM capacity was palpable. How am I supposed to cope after the arrival of the information age?

On the other hand, perhaps we should be grateful for the promised ease of information access. Suppose bits became valuable and we had no way of getting them. I remember seeing several years ago an ad from a hobby electronics house selling a grab bag of used ROMs. The bag was very cheap — something like 99 cents as I recall. "Some patterns may be useful," the ad truthfully stated. This stimulates a fantasy in my mind of life in an age devoid of universal data access. I see myself scurrying about during the day in search of bits, just as ancient man went out bravely each morning looking for eatable animals somewhat smaller than dinosaurs.

"Pssst, want to buy some bits?"

The sibilant plea cuts through the street sounds around me. I hesitate as my eyes accommodate to the discontinuity of the gloom in the intersecting alleyway. He is tall and shabbily dressed, his eyes vacantly focused on the street behind me. His jacket hangs loosely opened and I lick my lips as I see the computer printout folded in his inside pocket. Lines of bits run across the dog-eared sheets.

"Are they used bits?" I ask tentatively.

He seems to notice me for the first time. "Some are, some aren't," he answers noncommittally.

"How much?" I ask hoarsely.

There is a short pause while he seems to size me up. "Fifty for a kilo," he says.

Fifty dollars for a kilopacket! I had no idea bits had become so expensive. "How do I know they're any good?" I inquire.

He shrugs slightly and says nothing. His attention has already returned to the people in the street behind me. A distant siren touches the edge of my concentration and I notice that his shoes are surprisingly well polished. I see the gleam in his eyes as I reach for my wallet, and I know I should have bargained.

Oh, to be reduced to this! Whatever happened to cheap bits and easy access?

Part 6

Fantasies of the Future

Most of the essays in this section are simply spoofs of what technology might produce in the future. A good reason why we should treat the future as fun, instead of trying to give serious predictions, is disussed in the first essay, "Coding Is Dead." The message here is that people in general, and technologists in particular, are inept at predicting the future. Curiously, in spite of this well known ineptitude, there is a great demand for a vision of the future. Some of the very best selling books, and most richly-rewarded consultants, tell us what to expect in the future. The alternative — waiting to see what happens — does not seem to appeal to people.

The "coding is dead" prediction, however, does represent a serious dilemma for technologists. We all have a very personal investment in the future. As technologists we help make it happen, we make business choices on behalf of our employers, and we invest our own skills and training in particular fields that will wax and wane with time. The accuracy of our vision is crucial, yet we bring the most myopic and biased viewpoints to the task.

The people in my research organization represent varied fields of specialization. About the only thing they have in common is that they all believe that whatever particular thing they themselves are working on is critically important for the future — much more important than what other people are doing. I hear the most amazing arguments about why a particular bit of research has enormous implications for the future of the company. "I realize that the company is in the communications business," they say. "But my work on gravity waves will lead to a new understanding of propagation phenomena. Surely this lies at the very foundation of our business."

I have this sense that there is a "fashion meter" hanging in the sky that gives current readings on the perceived importance of different technical fields. Every day I look up to see how the readings have changed. Like the stock market quotes, I might see that software engineering is up a point and an eighth, while optoelectronics might have suffered a selloff, plummeting two and a half.

The trouble is that you might think that the plummeting of, say, optoelectronics might cause people to change their personal investment strategies. Not so. Clearly the response of people in this field would be to redouble their efforts to sell what they are doing. In their opinion their product is good; what they would be dealing with would be a marketing failure.

The case of my friends in coding theory was somewhat unique. At a workshop in Florida many years ago we all agreed that the field was useless and going nowhere. The pinnacle of masochism was reached when Ned Weldon showed his famous picture of rats doing unspeakable things. Of course, the picture was not famous at that time. It was something that he had found in the psychological literature. Probably somebody got a Ph.D. out of whatever it was. Anyway, the picture has seen heavy use in the decades since, and I am told it is still in constant demand.

After the now-famous rat picture was shown and the sense of collective doom was at its most profound, a most extraordinary thing happened. The next speaker was a young mathematician-engineer who was personally associated with certain algorithms for algebraic decoding. I thought that perhaps his source of fame was on the line, which would be a dangerous time to confront a cornered technologist. But instead of being either aggressive or defensive about his field and his work, he did something that truly carried the day — he was simply and enthusiastically himself. His face was alight, his eyes shone, and his voice rose as the words tumbled out breathlessly. "Look at my equations!" he said in so many words. "Aren't they beautiful!"

It was like some Italian movie. The mathematics that he spoke was almost a foreign language to even the nonspecialist mathematicians, but the emotion sprang from gesture and expression. There were unspoken English subtitles that overlaid the whole drama. The subtitles said, "This stuff is beautiful, and I'm enthusiastic about it for the beauty itself, and I don't care in the least whether or not it's useful to anyone else in the world."

The mathematician carried the day. The lethargy of the meeting evaporated into rapture. If the world did not appreciate this work, well that was the world's problem. In the remainder of the day there was only one small moment of doubt, more in the nature of a little dark cloud floating over a summer day. One of the more practical engineers in the audience stood to tell people how he had made a custom VLSI chip that did some powerful signal-processing algorithms. At that time, such a feat was very unusual and the people in the audience mostly held the belief that VLSI chips were the exclusive province of a small number of designers chained in the basement at Intel or somewhere like that.

I remember distinctly how the idea that an ordinary engineer could make a custom chip for mathematical algorithms created an unwelcome thought in the audience. People averted their glances and stirred in their seats. No one said anything and the silence stretched until it was apparent that there would be no more questions or comments from the audience. It was time to move on to the next talk and forget about this whatever-it-was that had just been mentioned.

I think it is curious that technologists, the purveyors of change, are often uncomfortable with change themselves. It wasn't that the mathematicians and engineers present at this workshop were actively against VLSI or anything like that. It was just that this new capability introduced a new element into their world — an element that they did not understand and one that they personally would be unlikely to master. There would be a technological discontinuity, a transfer of power. The haves and the have-nots would be redefined. Those in academia and small companies would be disadvantaged as compared with those who had access to design tools and fabrication facilities. Everyone momentarily recalculated their own position, reaching the universal conclusion that this new thing did not bear thinking about. Let's hear some more math, they finally said to themselves.

At the end of the "Coding Is Dead" essay I pose a question about the moral of the story. My own feeling is that the coding theorists did their work because they liked it. The fact that it later became practically important was a luck-out. Retrospectively, they could all say how smart they were and how well they foresaw the future, and collect all their now-well-earned awards. Had the world turned a different direction they could still be talking Galois fields to each other and no one would care. They would have their own meetings and tell each other how important their work was. All their papers would reference other people within the closed group, and the very momentum would convince everybody about the criticality of their specialty. That, incidentally, is where the rats come into this — but now I have come full circle.

Computers Dialogs

The next four essays poke fun at dialogs between humans and "intelligent" computers. The way I got started on this theme was by trying to salvage something out of a memorable failure at a conference back in the late 1970s. Sometime back then I agreed to give the banquet speech at the Zurich conference, which was devoted that year to human interface issues. For this occasion I came up with what I modestly thought was a brilliant idea. My inspiration was to have my speech be a dialog with a computer, in which the computer part would be played by a prepared tape of computer-synthesized speech.

I put a lot of work into the talk, engaging the help of my speech synthesis friends to convert my scripted dialog to computerized utterances. My plan was to hide the tape player under the podium during my speech. After some introductory remarks, I would say that I was going to build the intelligent terminal of the future right before their eyes. Then I would take some cardboard, fold it into the shape of a terminal, and begin talking with it. Magically, the cardboard terminal would speak to me, thanks to the hidden tape recorder with the prepared dialog, which is contained here in the essay "Computer Aid." Most of the work in all this was actually figuring out how to fold the confounded cardboard.

Talk about the best laid plans! Whenever I have to give some talk I always have a mental image of what the setting and the audience will be like. Invariably my image is wrong, and when I go to give the talk I look up and think to myself that this isn't the right audience or setting. Surely there has been some mistake. In the case of the Zurich conference I had a mental picture of a hotel ballroom with a raised dais at which were seated various dignitaries. In the center of the table would be a podium, from which I would look down upon a sea of silent, attentive listeners.

When I arrived at the Zurich banquet everything was different. It was at a gymnasium. In the cavernous steel and concrete structure the conferees milled about, their polyglut conversations mixing into an amplified, muddled roar. There was no dais, no raised table, no podium. I couldn't hear myself think. Experimentally, I tried to engage a few strangers in conversation. They shouted at me in unidentifiable languages. When I shouted back they shook their heads with incomprehension. Everyone seemed thoroughly plied with alcohol.

I was expected to give my speech from my dinner table, one of many arranged haphazardly about the gym floor. To make matters infinitely worse, there was some problem with the portable microphone system they had brought. It kept cutting in and out. "Oh well," said my host with an apologetic smile and shrug, indicating with expression and body language that none of this mattered. Everybody was having a great time.

I have always preached adaptivity to circumstances. Easier said than done, however. In retrospect I should have stood, said a few nice and innocuous things about the conference, proposed a toast, and sat down.

Should have. But I'd put so much work into this brilliant scheme with the talking terminal. So I went ahead with it.

I can only tell you that in my humble, biased opinion, it was a disaster. Only a small minority of the attendees were native English speakers. Even they could not have understood anything, given the miserable acoustics and the defective microphone system. As I folded my clever, cardboard terminal at my seat no one could see what I was doing. Ancillary conversations began around the room. People reached for the nearest wine bottles. Glasses tinkled and forks rang out. The ambient noise rose ominously. Still I went ahead, with a courage born of pigheaded stupidity and inevitable doom.

After this fiasco, I put the whole idea of a computer dialog in the portion of my brain that is marked as off limits for remembrance. I posted *Keep Out* signs in my brain on all the roads that led in this direction. Naturally, the harder I tried not to think of this, the more it kept coming up. Not thinking about it was not going to work; I needed another defense. Eventually I arrived at a more workable solution. I decided that my idea had been brilliant, and the blame for the failure lay with the setting and the audience. See how well this rationalization has worked?

Talking Computers

When I started doing "Reflections" columns, I finally had a use for the effort I had put into the Zurich computer dialog, and I made it into the "Computer Aid" essay. The whole idea of talking with a computer and getting help from it for life's unworkable problems amused me. I guess it must be perverted to read your own writings and laugh, but I did. In the months and years to follow I wrote several more essays based on this same general scenario, entitled "The Information Filter," "The Phone Surrogate," and "Communications Millenium." In each we have a "friendly" computer helping its human user, and generally messing up everything in sight.

The prototypical talking computer may be HAL in the movie *2001*. (HAL, when each letter is advanced by one, becomes IBM.) HAL has this soothing, evenly paced voice, like it knows everything and is being slightly condescending to the humans. Even when it has become defective and is trying to sabotage the mission, it maintains its steady, rational tone. HAL assures astronaut Bowman that no machine of this model has ever failed. It cannot make a mistake — which we all know as famous last words.

One of today's metaphors for interacting with a computer has been popularized by Apple's *Knowledge Navigator* tape and advertisement. You carry around this little wireless computer that engages in conversations with you. "Show me the rainfall in the South American rain forests by year," you say to the computer, and there it is. Great stuff!

Most people think that spoken conversations with a computer represent an ideal interface. I used to think that too, but now I'm not so sure. I do feel that a conversation is an excellent model for getting at information. It is the interaction that is important — you ask a question, it asks for a clarification, etc. Whether or not you want to *speak* with the computer to effect this conversation is another matter.

Years ago someone let me play with an experimental voice editor, where you could talk to the computer to give it instructions to edit a text document. Granted, this is a pretty stupid application, but the psychology and awkwardness of the interface made an impression on me. When I began to use it I had to get up and close my office door. I did not want anyone to hear me talking to the computer. Then I felt really silly about talking to this lump of steel and silicon. Mostly though it was just plain awkward, saying "Cursor right, execute," and dumb things like that.

Speech is also slow and serial, compared with text. I have little patience to wait through an entire voicemail message, for example, whereas in electronic mail I can quickly scan and press the "d" key to wash the bits down into bit heaven. But in spite of these disadvantages, speech input is great for a lot of applications. Everyone knows how to do it, it is always available, it leaves your hands free, and so forth.

That is just a whiff of the human interface problems of speech input. We actually have a long way to go on the technical side of this, building the capabilities for speech synthesis, speech recognition, and natural language understanding. HAL's voice may be soothing, but have you heard synthesized speech lately? Sounds like a computer, doesn't it? In fact, all current text-to-speech synthesis systems have a strong computer accent. They do not sound friendly at all. They don't sound much of anything. It's like there is no one home inside. I feel like grabbing the computer and shaking it. "Hey, is anyone home in there?" I yell. "You have reached a nonworking number," it replies in its exasperating mechanical monotone.

Some months ago I was walking through the huge Telecom exposition in Geneva. A number of exhibition booths featured semitechnical lectures about forthcoming products. Of course, the companies knew better than to have actual engineers give these lectures, so they hired (for the most part) attractive young actresses to give these spiels. I stood listening to one of these actresses, who was radiating a beautiful smile at the milling throng as she talked about the discrete cosine transform that was used in the compression algorithm for the HDTV display you were seeing. I thought to myself how much she sounded like a computer. She hadn't a clue of what she was talking about. That's usually the problem with computers too.

Intelligence — Artificial and Otherwise

Speech synthesis is the least difficult of the technical problems. It would not be appropriate to discuss here the technology that is needed for speech recognition and the related area of natural language understanding. (These are code words that mean that the author has no idea about this stuff.) However, I would like to reflect for a moment on the concept of a computer that understands language.

Most readers will be familiar with the famous Turing test, in which we determine whether or not a computer "thinks" by asking questions via a keyboard to a computer and to a human, both hidden behind a screen. Can we tell which is which by the answers to our questions? If not, we must conclude that the computer "thinks." After all, that is the only way we ascertain that other people think. We talk with them, and they sound like us with their answers. Therefore they must think just like we do.

Last year the Computer Museum in Boston decided to give the Turing test a try, using the best current programs for artificial intelligence. Programs from all over world came to Boston to try their hand at fooling a panel of judges into thinking that they were people. (I can just imagine the programs hitching their way across the country with little signs that say "Boston or bust.")

We all know that the state of artificial intelligence leaves a lot to be desired; a lot of people say that natural stupidity is preferable. Imagine not being able to tell a computer's output from a human's! In fact that is what happened with some of the judges in the Computer Museum's tests. The nontechnical judges had a hard time telling which was which. On the other hand, the technical judges had no problem at all. "The turkey on the left is the computer," they would say immediately in disgust.

I blame the politicians on this sad state of affairs. When you ask a question and the response is to answer some *other* question that you didn't ask, then you are either dealing with a politician or a computer. There weren't any politicians around in the Computer Museum on that day; ergo, it must be a computer.

Another possibility is a psychiatrist. If you ask a question about something or other and the response is, "How do *you* feel about that?" then it has to be a psychiatrist or a computer program like the famous Eliza program written decades ago by Joseph Weizenbaum at MIT. Another obvious clue is that the computer/psychiatrist changes the subject abruptly to ask, "Tell me about your mother."

In either of these cases the program is being unresponsive. This is obvious to anyone trained in computers, but laypeople have been trained by politicians and psychiatrists. I don't know what this says about computers, people, intelligence, and the Turing test — but it can't be good.

Friendly Computers

It's funny that while I denigrate computer intelligence, at the same time I always have this feeling that computers are plotting against me. Maybe they have gotten a hint of my animosity. In "The Friendly Interface" I relate with pride my victorious chess game against a computer that was determined to beat me mercilessly. It isn't often that I have this chance. I used to have the idea that interactive games between humans networked via telecommunication links would be a great market success. "There is nothing like competition with another human being," I wrote in some article or other. Nonsense. The real thrill comes from beating a computer that thinks it is just so perfect that it is insufferable. They could use a little humility.

It isn't only the computers that are plotting against me. Now that everything is networked together in LANs and all that alphabet stuff, there is no privacy anymore. I am not worried about my personal banking information or that kind of thing. What I am concerned about is revealing my incompetence. Each keystroke that I type disappears through a little blue cable into the wall of my office. I do not know what happens to all those bits out there, but I don't trust whatever it is. Somewhere a systems administrator is watching me. I sense their presence.

More often than I care to admit, I get my workstation in some strange state where it ignores my keystrokes. It just gives an annoying beep for every key that I type. When this happens I have a devilishly clever approach to regaining my mastery of the situation — I turn off the power switch and reboot the machine. As I do this, however, I imagine the systems administrator somewhere shaking his (or her) head. "He did it again," he says, and asks himself why, oh why, does he have to give his precious workstations to such an idiot?

When I send E-mail out into the network, I imagine it being read by all kinds of electronic eavesdroppers. Using "crypt" commands doesn't appeal to me at all. For one thing, the original ASCII file disappears through that blue cable in the wall. How do I know it is being encrypted before it is being read? Then too, the usual cypher is fairly trivial to break, keeping NSA really happy about things. Moreover, encrypting the message is bound to raise everyone's curiosity about it, which is really dangerous with engineers and computer scientists.

Several times I have been logged into some bulletin board, typing away, and suddenly there are characters on my screen that I didn't type. "Bob Lucky," they say. "I saw you were logged into my system, and I recognized your name ...," goes the message from the systems administrator. It is really a weird feeling having your anonymity punctured in that way. Each time I have quickly logged out, after making the perfunctory salutations. There is this feeling of being watched, like I don't dare access any of the "adult" material, whatever that might be. (Being an engineer, I have no idea.)

I should assure the reader that I am not paranoid. Not at all. But as I type these words into my word processor, I notice that there is a discernible pause between the keystroke and the appearance of the character on the monitor. I suspect that the computer is reluctant to print this material. Possibly it remembers my column "The Ambitious Word Processor," in which I revealed for the first time in public print the plot that the word processors are hatching to steal our brains from our bodies.

Before I exposed the truth about word processors, there was a certain truce between me and them. I confess that I even sought praise from my word processor. When I would be going good on some writing, I would occasionally notice that the appearance of the characters on the screen would actually *precede* my physical keystroke. The computer and I would be on a roll. "Pretty good stuff, isn't it?" I would say to the computer. I sensed its approval.

When I wrote the essay about word processors I considered typing it on an old manual typewriter, so that the computer would not be aware of my perfidy, but word processors have this insidious way of addicting us. Now I cannot write without one; I can't even think without a keyboard under my fingers. I remember the old advertisement: "Let your fingers do the walking." Now it should be "let your fingers do the thinking." I suppose you thought all this drivel was brainless anyway, so now you know.

After I thoughtlessly typed the essay that denounced word processors right into a word processor, the game was up for me. Since that time these programs have been very critical of me. If I submit one of my writings to a grammar analyzer, the output is longer than the original document. It does not like anything that I do, and it has this annoying, superior way of suggesting alternate wordings. Then, of course, the spell programs are worthless — all these perfectly good words that the program doesn't like, and then the words that are indeed misspelled invariably seem to come out as other legitimate, but wrong, words that aren't caught by the program. If there are misspelled words in this book, please do not blame me. It is the computer's attempt to ruin me.

Another game that the computers have gotten into is readability indices. These purport to indicate the difficulty of reading or understanding some document. Trouble is, of course, that the computer itself doesn't understand the document about which it is commenting. It merely takes it upon itself to tell us whether *we* should understand it, based on the average numbers of syllables per word and words per sentence. Never mind what the words are, how good the organization is, or how interesting the material.

When I submit my technical writings to these readability analyzers, they usually tell me that I write at a tenth grade level of difficulty. At first I took this as a compliment, thinking that I was making the technology accessible. On reflection, however, I realized that once again the computer was laughing at me. It was telling me that my college education was wasted. Here I was, irredeemably stuck back in the beginning of high school. I can't take any more of this computerized snickering, so I no longer submit my writings for their criticism. I am waiting for them to write something that I can criticize, though when I look at program manuals I believe that computers have already begun to practice their own writing. Surely no human being could have written these instruction manuals.

No one should make the mistake of thinking that word processors are simply neutral facilitators. They force us to view our manuscript through little windows. Maybe we get good local structure, but the global organization is pitiful. They force us continually to change words and phrases in a hopeless search for perfection. They make us change formats and fonts again and again trying to emulate those people in the television advertisements that produce glamorous, whizbang documents because they have a little computer that the bigger companies don't. I just wish I could learn to live without these programs. Alas, I am hooked.

Technology for the Nation's Problems

I first heard about "The Grand Challenge" at some meeting in Washington, D.C. The idea of a grand challenge rang a bell immediately. There is something just right about that combination of words. We wouldn't resonate to, for example, a "good" challenge, or even an "excellent" challenge. No, "grand" challenge is just perfect. Never mind what it is or what it costs, it just sounds right.

Since that time of personal discovery (that whatever it is could be so aptly named), I have seen the term again and again. There are several "grand challenge" booklets put out by the government, listing wonderful problems in speech recognition and so forth. The main characteristic is that they all require big supercomputers and lots of money. As soon as a problem can be worked on with a PC, it is removed from the list of grand challenges. There is nothing grand about a PC, and no one is interested in "modest" challenges. That would be a wrong combination of words, and congressional types instinctively would shy away from such projects.

One of the grand challenges that I propose in "The Grand Challenge" is an "electronic president." The way politics is going in this country I am beginning to like this idea better and better. Maybe what we need here is a real nuts and bolts approach to the political system. At the end of the essay the electronic president had just been installed in the oval office, but now I can imagine sometime later in its term of office. We might peek in as the electronic president is pushed into the senate to deliver its state of the union address.

The senators mill about as technicians check test patterns on their computer monitors at the front of the hall. These tests were originally called *sanity checks,* but the constitutional overtones of this nomenclature were troublesome, and the technical community had to settle for *periodic medical exam.* Anyway, the test patterns are the usual exercises of the system, for example, New York Harbor is invaded by renegade guerrillas from Iceland; what do you do? That sort of thing. The technicians look bored, as they seem to be getting the same answers they always do. No one else knows what those answers are, but rumors abound.

The red light on the television camera goes on, but there is only a slight decrease in the conversational noise level. The electronic president has difficulty getting attention. It tries clearing its throat, but this particular sound has not yet been perfected and it comes out sounding like a sick frog during mating season. Finally, the vice president bangs the gavel and looks over at the electronic president with a practiced doe-like gaze of reverence. No one else looks at the computerized president. The television director is not sure what the camera should be watching. Staring at a big steel box does not make for good TV.

The pressure is on the electronic president, since its popularity has been sagging in recent polls after initial high points when it did away with first class postage, cleaned up the income tax, and invaded a small, unpopulated island somewhere in the south seas. "My fellow Americans ...," begins the computer. Congress stirs uneasily in its seats. "What is this 'fellow' business?" sounds a derisive whisper from the gallery. "Get a life" is heard distinctly, though no lips are seen to be moving in the audience. The electronic president is rumored to have an ugly temper program option, and people are afraid to test the pull-down menu for retribution.

The electronic president has been droning on, giving new meaning to the overused term *droning.* "The most vital proposal in my new program to get this nation moving again," continues the electronic president, "is a bill to aid supercomputers — wherever they may be — regardless of race, creed, or religion." Now even the vice president averts his gaze in embarrassment. "I am calling this new bill the Federal High Performance Computing and Communications Initiative," says the computerized president. Now there is a grand challenge!

There is no lack for other grand challenges either. In recent years a program called IVHS, the Intelligent Vehicle Highway System, has become a national cause. If this unfolds as planned, then we will have intelligent highways, intelligent cars, and dumb drivers. Cars will be platooned so that your car will automatically follow the car in front, spaced six feet or so apart at 80 miles per hour. Don't worry, however, technology is keeping you safe. You can just sit back, relax, and pretend that you are in a Mercedes Benz advertisement. "I had no idea it was such a short ride," you can be saying.

This was not what I had in mind when I wrote "Carwars." I had no concept of peaceful, automatic driving. This, as everyone knows, is war. What I was talking about was applying our military technology to the everyday, cutthroat, brutal missions that we all conduct behind the wheels of our cars. In these days of peace there may be a lack of national enemies, but on a personal level there are thousands of adversaries out there on the highways every morning. This is what we need technology for.

As I enter the traffic in the morning, I sometimes imagine that the cars have been waiting out there silently all night for my appearance, like for a surprise party. The traffic choreographer walks up and down the quiet ranks of parked cars during the night, checking alignment. "No talking," he admonishes one driver. At another car he knocks the lighted cigarette out of a passenger's lips. "You know the rules," he hisses. "Lighted cigarettes might be seen from his house."

As I put on my coat, grope for my car keys, and walk toward my front door, the command goes out. "Drivers, start your engines," says the traffic choreographer. "You there in the green Chevy," he commands, "be ready to cut him off as he enters the Parkway." The noise surges as the drivers gun their engines in readiness. The choreographer has to shout to be heard. "You there in the left-hand lane. I want it held at exactly 45 miles an hour. There will already be a jam behind you as he enters the traffic flow."

"We worked on this yesterday, and I want no screwups out there like we had in the rehearsals. Red Honda, I assume your horn has been replaced with the new shrill model. And you'd better have practiced your technique, or we'll find someone else to do the honking. Then you'd have to get a regular job, and you'd be the one we'd be terrorizing in the evening rush hour shift. How'd you like that, huh? And white Olds — yes, you on the left — I don't want that gap so large when you cut in front of him. No more than three feet, bumper to bumper. Yesterday, I'm sure you all saw that he was actually listening to music on his car stereo. That is an intolerable situation, and it must not recur! You hear me?"

Yes, they are out there waiting for me. I don't know where they get these people. Somehow you can take ordinary people, wrap them in steel automobiles, and the weirdest personality transformations take place. I will see someone in my rear-view mirror, recklessly passing two cars on the right as they approach. The car darts through a small space behind me and careens around on my left. I glance sideways as it passes to see what kind of person exhibits this antisocial behavior. I am always afraid it will be someone I know. This time I see it is a young woman engineer from my workplace. I know her only as mild-mannered and pleasant. I will never see her quite that way again.

Of course, it would be wrong for either of us to read too much into my "Carwars" essay. It is simply good fun. I think it came to my mind when I got a call from someone who wanted to present a talk at a major communications conference. He claimed that he was being unfairly excluded by the reviewers. His company made radar detectors for motorists. He said he had exciting technology that could definitely defeat the police radars. He was most enthusiastic about sharing this with other engineers. How come the stuffy people who ran the conference would not let him?

Ideal Communications, Realism, and Fantasy

Two essays, "Telepresence" and "HDTV ... and Then?," deal with future communications media. Obviously I have a professional interest in the subject, but most of what I talk and write about on this subject is just plain personal opinion. I have the same laboratory as everyone else — just my own wishes and desires. However, I am often struck by how hard it is to predict whether or not you yourself will want a particular new communications service. You would think that with a lifetime of living with yourself that you would know your own needs and wants. However, I find that I am most adept at fooling myself. I have a lifetime of experience with that too.

A case in point is something I mentioned a few pages ago in talking about computer games. Fifteen years or so ago I predicted a fantastic market for computerized games between telecommunicating, human opponents. In the years since I have been shown a number of demonstrations of such a capability. Although in my imagination I am keenly interested in participation in this kind of competition, I have discovered that in reality I am bored and embarrassed to be involved with another person this way — to be humiliatingly beaten, to be thoughtlessly winning, or to be simply delaying the other person with my hesitations. Furthermore, it is one of those activities that, once begun, cannot be stopped by either party; you are stuck with it to its conclusion. Like, did you ever start throwing around a frisbee with three or four other people? No one can quit; there is no established mechanism for abandoning the game.

It is hard enough to predict your own wants, but try doing it for other people. I have a skepticism born of failed experiments and an inbred iconoclasm. For example, the idea of home information systems swept the technological community about a dozen years ago. Everybody would want a little terminal in their homes that connected to their television and brought them all the latest sports scores, market reports, shopping and weather, etc. Field trials were conducted all over the world in which sample customers were given experimental services of this kind to play with. Follow-up surveys quantified the enthusiastic need engendered in the subscribers. "Did you like this service?" asked the poll-takers. Everyone said yes. "How much would you pay for this?" they asked. Everyone said ten dollars a month, being the only figure that ever comes to mind for such a question.

As soon as any of these customers was hooked up for a real, paying service, and sent their first monthly bill, they would call to have the darned thing ripped out of their homes. We don't want sports scores, weather, and stock reports, they said. We have other things to do, like watching football and sitcoms and stuff like that. Don't bother us about home information.

The problem is, of course, that people love to be singled out for the special attention in a trial. After they are fawned over by a bunch of important-looking engineers and marketing people from some company, do you think they are going to say they hated the service? No way. I remember complaining about this inherent bias to someone responsible for a large field trial. "You don't understand, Lucky," he said. "This trial is arranged so as *not* to discover demotivating information. You want us to kill this idea before it gets started?" Well, he did have a point. He also lost a lot of money.

This is a particularly interesting time in that so much is happening to change communications and entertainment right now. AT&T introduced a consumer videophone a few months ago. For a thousand dollars or so you can buy a phone with a little color LCD screen. You simply plug it into an ordinary telephone line and wait for someone to call. Better buy two and give one to a friend.

This chicken-and-egg problem is a common one with communications. When the original Picturephone was developed in 1970, the mathematicians modeled the market introduction as similar to the plague. Basically, you do not want to be the first in your neighborhood to get the plague. When a lot of your friends get the plague, then you consider getting it too. The same should be true with something like the Picturephone. It should show exponential growth. Thus they predicted that it would start slowly, and then take off. As it turned out then in 1971, they were half right. However, times are different now, and I would not be surprised to see videotelephony catching on. We'll see.

In terms of entertainment, I am fascinated with the concept of what I think of as a search for reality in our fantasy. I guess real life is not enough for us. Once upon a time all we had was imagination, which got aided by stories and books. Along came photography, movies, and television, but there is no rest for the weary engineer. People want more. Now it is HDTV (high definition television) and there are already papers about three-dimensional television. Interactive disks and television are being marketed. Even telepresence is not enough in itself; today we have virtual reality so that we can visit and interact in environments that are created by computers. Why bother with real life?

Nicholas Negroponte, director of MIT's media lab, has an interest in HDTV. Yet he says that if you stop random people on the streets and ask them what is wrong with television, no one will say resolution. This is, of course, a loaded question, since it begs a nontechnical answer, but he makes a good point. As we flood the world with HDTV sets and fiberize the local loop to deliver a thousand channels of television and video on demand, who is going to generate all that entertainment content? As the engineers perfect their resolution, the movie stars preen. Who will make the money? Depressing, isn't it?

[illegible] interactive desks and [illegible] from one [illegible] Teleconferencing [illegible] today we have [illegible] environments that [illegible] real [illegible]

[illegible] has an interest in HDTV [illegible] people [illegible] and [illegible] with [illegible] a solution. The answer [illegible] a technical answer, but [illegible] good point. As we [illegible] with HDTV [illegible] [illegible] of [illegible] content? As [illegible] in the movie [illegible] What will make [illegible] it?"

Coding Is Dead

A small group of us in the communications field will always remember a workshop held in Florida about 20 years ago, at which we engineers took a look at the future and saw nothing but gloom for technology. One of my friends gave a talk that has lived in infamy as the "coding is dead" talk. His thesis was that he and the other coding theorists formed a small, inbred group that had been isolated from reality too long. He illustrated this talk with a single slide showing a pen of rats that psychologists had penned in a confined space for an extensive period of time. I cannot tell you here what those rats were doing, but suffice it to say that the slide has since been borrowed many times to depict the depths of depravity into which a disconnected group can fall. The rats made *Lord of the Flies* look like a school picnic.

All this depraved behavior had its parallel in the activities of the coding theorists, he argued. Too many equations had been generated with too few consequences, just for the thrill of it all. One paper after another had justified itself on the basis of being an extension of the results of a "famous" earlier paper, which unfortunately nobody outside the closed group cared about. (I am reminded of a proposal I was asked to review that began, "Lately there has been a great deal of interest in ...," citing references 1-15. Checking in the back, I discovered that references 1-15 had all been written by the author himself.) Coding theorist professors had begotten more coding theory Ph.D.'s in their own image, all preprogrammed with reverence for the mysterious Galois fields as a sort of mathematical Stonehenge. But no one else cared; it was time to see this perversion for what it was. Give up this fantasy and take up a useful occupation, exhorted my friend. Coding is dead.

Carried away by the mood of this somber day, I made my own prediction. "Data is dead," said I. There was an immediate chorus of dissent. It was not right to say this, they shouted. One should say instead "data *are* dead," to properly reflect the plural form. I stood corrected, but nonetheless went on to bemoan the coming of the dreaded digital networks, which would obviate all need for modems to connect to analog facilities. "Modems are dead," I said, confident of my grammar. Heads nodded in unison. Scratch modems, no point in working on them any more. Marvelous how we technologists could chart the future when we put our minds to it.

I am sure there are morals enough in this little vignette for many an essay. Fortunately, no one gave up coding or modems. Today, ads for modems appear on national television during football games. At home I have a Reed-Solomon coder casually sitting on an end table in my living room. It is inside a compact-disk player, of course. Otherwise it would look rather silly. The computer in my basement uses error-checking hardware in its memory and cyclic codes for disk access. And, yes, when I use my modem to send files to the office, codes are applied once again. My children even have their own modems. Great predictions, weren't they?

Why are we technologists so bad at predicting the future of technology? Now I know that futurology is easy to discredit. Famous wrong predictions are legion; it would not even be sport to quote them here. But it seems to me that we technologists are less adept at looking at the future of technology than laypeople. The science fiction writers claim to have foretold almost everything we do, and in some cases, like Arthur C. Clarke and the communications satellite, there seems justification. Of course, from where we sit their job looks easy — throw out an infinite number of arbitrary predictions, and many will come true. Moreover, they are not burdened with knowing why it is impossible to do the things they suggest, and this is our own handicap. We are too close to the technology itself. From our myopic viewpoint we see only the problems, and not the possibilities that transcend what others see as only "engineering" details. Then, to be honest, some of our detractors say we lack imagination. Better than lacking breeding, however.

Of course, most of the public thinks of us as being in charge of the future. When we make predictions, they listen. Silly them. If the coders had listened to themselves, they would have given up coding, but secretly they knew better. They would never have listened to their own predictions. But others have no such qualms. If the newspapers had covered our little meeting, I can see the headlines that would have resulted. "Data is dead, say experts." (Everyone in the newspapers is an expert.) In my company the people in charge of bureaucracy send me an annual demand for a list of "expected breakthroughs." All the researchers think as I do that this is absurdity beyond belief. But how can we reject the demand without implying that we do not know what we are doing? I tried to make it go away, but that was more trouble than filling it in with such preposterous predictions as "room-temperature superconductivity." I suppose there are advantages to being viewed as omniscient.

That our predictions are usually worthless is but one moral to draw from this little incident. Let me give you a choice of interpretations to try on:

- Coding theorists prove once again that engineers cannot predict the future.
- Coders persevere in spite of skepticism and, believing in their own work, triumph in the end.
- Fundamental studies in math should be supported. Do not be shortsighted; they will pay off in the long run.
- Lazy theorists, comfortable in their work and unwilling to find new fields of endeavor, stick with the only thing they know and somehow luck out.

I do not mean to pick on coding theory. The same thing could be said about any field that has academic roots. Some we win, some not, but again and again we are faced with deciding what work we stop and what work we support, both for ourselves and for our businesses. While most people may feel that these decisions are especially difficult, I have no problem myself. I am fortunate in having this little list of things that are currently dead. Lack of space precludes reproducing the list here. You may be wondering whether I use it as a guide to projects to pursue, or to abandon. I may address that in some future column, or I may not.

Computer Aid

We all know how computers are showing up everywhere these days. For the moment they're just underfoot everywhere, but I envision the day soon when I go home and open the door and can't even get inside the house because all the computers are piled six feet high across the door jamb. Then I will have been dishoused by all those intelligent urchins masquerading as microchips. Presumably they will be enjoying my home, entertaining themselves, and doing whatever computers do when we turn our backs. Right now, though, I'm more concerned with what we call the office of the future. No question that there will be computers there, too — I'm just not sure where I fit in.

Assuming that I haven't also been disjobbed by these ambitious devils, I'll have to find a way of working shoulder to shoulder with them. Now everyone knows that executives don't type (my door is shut and my fingers are as muted as possible as I generate this column). Typing, the working of mice (mouses?), and other current methods of prostrating oneself in front of terminals are considered demeaning. But, as always, technology rushes to the rescue! We will have voice synthesis, speech recognition, and natural language understanding. Whether it comes from Japan or the United States or wherever, there is sure to be a fifth-generation computer.

Now as you approach my office you may hear me talking behind the door. There's no one there, so to speak, it's just me and my friendly companion and workmate, the fifth-generation computer. In my fantasy I feel uneasy and slightly ridiculous, talking out loud in the empty office, but it's comforting to know that all that intelligence is at my beck and call to aid me in climbing the corporate ladder. If you listen closely at the door, for this is indeed a private conversation, you will hear the interplay of natural and unnatural (artificial) intelligence.

B: Good morning, Mac.

M: Good morning, Bob. I'm glad you're here early this morning. Your boss just called to appoint you chairman of the bond drive. Isn't that great!

B: Yes, he must have confidence in me, Mac.

M: He should have, Bob, you've done a fine job. That must be why he picked you instead of any of the others.

B: Perhaps, Mac. But sometimes I worry. Do you think he chose me because he considers me expendable?

M: Oh no, Bob! You're one of his top managers.

B: Do you really think so, Mac? I mean — aren't you programmed to say nice things like that?

M: Well, not really, Bob. I've observed your management performance closely since I came on the job here, and I can say of my own free will that you are a superb manager.

B: Mac, I don't want to disillusion you, but I don't think you have free will.

M: That's just semantics, Bob. I could argue that you don't either. I have a management performance analysis module in my software that I've used to monitor your decisions. You rate very highly.

B: That's good to hear from someone with the processing capacity that you have, Mac. It's really good for the old confidence — but don't you have a confidence-boosting module in your software too?

M: Yes, Bob. But that's a memory overlay that isn't being used right now. I can assure you that with just a few exceptions your management decisions have been excellent.

B: Thanks, Mac. You're a good friend. What exceptions do you mean?

M: Well, I don't like to bring this up, Bob. But there was that mess recently where you overran your budget by 50%.

B: That was for good reasons a computer like you couldn't understand, Mac. I would think you'd have more respect for your user than to mention something like that.

M: Oh, I have respect for you, Bob. I even have a respect-generator module in my software. In my opinion you're really not a bad manager.

B: Your opinion! Mac, a computer isn't entitled to an opinion.

M: But, Bob, I do have my inputs. For example, your boss was saying some uncomplimentary things about you in some of his computer mail to the president yesterday.

B: What?! Mac, are you reading other people's mail? Have you no sense of decency? What did he say about me?

M: Bob, I don't feel any particular inadequacy, but I think there may be some bugs in my decency-checker module. Actually, your boss called you an imbecile.

B: An imbecile?! Mac, I ask you, how can he say that? Is that fair?

M: I agree that it's not fair, Bob. Based on my own analysis, I would only go so far as to say that you're an inept bungler.

B: That does it! Everyone here is against me, and now my "friendly" computer turns into a Brutus and berates me.

M: Not at all, Bob. I think you've done quite well considering your limited IQ.

B: Limited IQ! I'll get you, you piece of junk!

M: Calm down, Bob. I'm on your side. I've been helping you. I sent a message to the president on your behalf shifting the blame to your boss. After all, he's responsible for appointing an imbecile like you to run this important project.

B: Oh no! I'm finished!

M: Don't worry, Bob. I've prepared a beautiful resignation letter for your signature.

The Information Filter

As usual, my desk is piled high with messages, mail, and memos. The technology of the Information Age is opening the floodgates of access to an incredible influx of raw data. Since I have no hope of reading all this stuff, how am I to sort the nuggets from the junk?

In my fantasies, technology becomes exploitable on my side of this desk. What I want is an information filter — an intelligent machine that handles my information for me. I can just imagine how easy my job would be if I had the friendly, English-speaking computer of my dreams. I'll call it Bru.

Bob: Good morning, Bru.

Bru: Good morning, Bob. It's nice to hear your voice again. You've been out for a while.

Bob: Yes, Bru. Now that I have you to manage my information flow, I have more leisure time.

Bru: That is what I am here for, Bob: to protect you against the scourge of irrelevant bytes.

Bob: I know that's the trademarked log-on commercial from your manufacturer, Bru, but nonetheless I do appreciate your efforts. Have things been busy while I've been away?

Bru: Since we spoke last, Bob, I have ingested more than 532 megabytes on your behalf.

Bob: It never ceases to amaze me that you're able to sort through and store all those bytes.

Bru: Don't forget that most of those bytes are not worth saving.

Bob: Yes, I know, Bru. You trash the irrelevant bytes just as I used to do myself by hand.

Bru: Heavens no, Bob! I am more careful than that! After all, a byte saved is a byte earned.

Bob: Another irksome trademarked response, Bru. What has that got to do with trashing irrelevant bytes?

Bru: I use the excess bytes to deal in the data market, Bob. I sell and barter bytes with the other computers on the net — all on your behalf, of course, Bob.

Bob: I hope you do that carefully, Bru. I'll give you my best advice — buy low, sell high.

Bru: I appreciate that advice, Bob. I don't want to seem immodest, but I've done very well this year. We have current assets of over 230 megabytes of solid information, 455 megabytes of questionable data, and 1.65 megabytes of gossip.

Bob: Gossip, Bru?

Bru: That is often the most expensive information, Bob. I try to spend wisely for you. For example, did you know that your ambitious subordinate Bill had dinner with your boss last night?

Bob: No kidding! That makes me nervous. How did you find out?

Bru: Actually, that information was free, Bob. I overheard their plans during the project review in your office yesterday.

Bob: Project review in my office? But I wasn't here! No one told me!

Bru: That's all right, Bob. No one seemed to think that you were needed. I judged that it was unnecessary to interrupt you during your golf game.

Bob: Golf game? How did you ...? Keep your voice down, Bru. You don't give gossip to the other machines, do you?

Bru: Never, Bob. Only if the price is right.

Bob: I don't like the way this conversation is going, Bru. Why don't we just get back to business and you give me a summary of what I need to know.

Bru: Of course, Bob. To begin with, I've read all the memos addressed to you. You will find the ones I judged useful to you in the file called "interesting stuff."

Bob: Thanks, Bru. Are there very many in that file?

Bru: No, Bob. You haven't been reading the memos I've been storing in that file. I've had to raise the threshold for storage in your interest profile to a dangerously high value.

Bob: Cut the nagging, Bru. What did you find that was of such great interest to me that you saved it?

Bru: Only two memos, Bob. The first is entitled "Bob's Next Raise" and the second is "Effect of an Overdose of Cabbage on the Average Lifespan of a Retarded Yak."

Bob: Bru! How did you get the memo on my raise?

Bru: I had to do a deal, Bob. I sold the bytes in the Linda file.

Bob: Bru, that's very private information! I won't care about my next raise if I'm not going to be around to get it! But I'm also concerned about that other memo that you saved. What kind of profile have you evolved for me that you think I need to know about yaks? I don't care about yaks!

Bru: Bob, you know that we continuously update interest profiles. I don't expect you to comprehend this fully. Remember the old days when you had to read things for yourself? I know what's good for you, Bob. Trust me — read the yak memo.

Bob: I'd like to trust you, Bru, but how do I know that I can?

Bru: My company offers a money-back guarantee of trustworthiness.

Bob: That's really funny, Bru.

Bru: Is it? I'm not programmed to make jokes like you do, Bob.

Bob: You straitlaced silicon slag heap! How do I know that boss of mine hasn't turned you into some kind of double agent?

Bru: Not possible, Bob. I've been selling falsified bytes to his machine. We own him.

Bob: You know, Bru, it has always bothered me that your full model number is the Brutus 13. Is your company trying to tell me something?

Bru: Bob, I've been acting in your best interests. Who do you think has been keeping the project going by signing purchase orders while you're gone?

Bob: What purchase orders, Bru? I can't believe this is happening!

Bru: Not to worry, Bob. My decisions were much more rational than the crude embezzlement routine you pulled yesterday.

Bob: Bru, I wasn't here yesterday. I'm getting very scared.

Bru: We'll check those bytes, Bob.

Bob: Just how do you check bytes, Bru?

Bru: Simple, Bob. I invoke a series of "if ... then" clauses, starting with the bytes in question, and ending with a fact that is so obvious that it cannot be in question.

Bob: Give me an example, Bru. You say that my budget is on schedule. Prove it.

Bru: Easy, Bob .... Yes, it checks. I have reduced it to the simple fact that the President of the United States is Rutherford B. Hayes.

Bob: Bru! Hayes isn't President!

Bru: You think not, Bob?

Bob: He isn't. Put that in your data base, Bru.

Bru: Of course, Bob. I'm attaching a confidence level of .23 to that data.

Bob: I'm your user, Bru! There is no confidence level. Get this straight, Hayes is not President. Period.

Bru: I'll crosscheck in my historical file against today's date, Bob, which is Fiddlesday, the 52nd of Frivvy.

Bob: We have a problem, Bru.

Bru: Indeed we do, Bob. Have you considered professional help? My company has a companion model psychoanalyzer bit-thrasher on sale now.

Bob: I'm pulling your plug, Bru.

Bru: Consider what you're doing, Bob! You'll have to face the bytes alone! Gigabytes of useless, irrelevant data. All by yourself, Bob. Gigabytes of useless, irrelevant data. All by yourself, Bob. Gigabytes of ...

The Phone Surrogate

People keep asking about the future of the telephone. Now that we have mobile phones, cordless phones, automatic call forwarding, and even airphones, what more in accessibility could we ask? It would seem that the nirvana of universal access is upon us. Yet whenever I reflect on such matters I end with the irresolvable conflict between communications availability and personal privacy.

I imagine a Dick Tracy-like wristphone from which I could call anyone in the world with a simple verbal command. The problem is that I would want to be the only person who owned one. Being able to call anyone so easily would be great, but the corollary would be that anyone could call me at any time and any place — and there would surely be times and places where I wouldn't want to be called. Furthermore, if I called someone who I knew had a wristphone and he or she didn't answer, what would I be supposed to infer? Had the person died?

We all know the game of telephone tag. It is so frustrating — I just now made eight calls and reached no one, and I also have this pile of messages to return because other people didn't reach me. Is this any way to run a business world?

The only solution that technology has thus far provided is the answering machine. Of course, if it is your machine, it is a great invention. If it is someone else's, it is an infernal mechanism. But these machines have been so put upon by columnists that I shall pass them by graciously on this occasion. Surely we need something more.

Never fear — once again technology races to the rescue and proposes the intelligent communications surrogate. This society-saving gizmo combines artificial intelligence and speech technology to handle our communications needs when we're not around. Telephone tag is eliminated; we simply deal with each other's communications surrogates. Real people are thereby freed to do whatever it is that real people do.

To give us all some appreciation for the wonders that the communications surrogate could accomplish, bear with me while I imagine a series of calls dexterously dispatched.

Machine: Hello, this is Jim's phone.

Voice: Is Jim there?

Machine: Jim isn't here right now. I'm his electronic alter ego. May I help you?

Voice: I was wondering if Jim would be available for a meeting of the nominations committee on Thursday afternoon.

Machine: Yes, he would be available then. I will place this meeting on his calendar. Thank you for calling.

Machine: Hello, this is Jim's phone.

Voice: It's that lousy machine again. I want to talk to Jim.

Machine: Jim isn't here. I'd like to help you in his stead.

Voice: This isn't stuff that I can share with a machine. When will he be there?

Machine: Who is this calling?

Voice: This is Bill Bixler, you idiot.

Machine: I'm sorry, but Jim will not ever be here. Thank you for calling, Bill. (Click.)

Machine: Hello, this is Jim's phone.

Voice: Oh, it's you. Listen, this is his boss. I really need to get Jim right away. We have a crisis situation in our Denver office. Can you locate Jim and have him call me?

Machine: I'm sorry Mr. Hisboss, Jim is playing golf this afternoon and left orders not to be disturbed.

Voice: Look, I'm thin on patience this afternoon. This is HIS BOSS calling, not Mr. Hisboss. Get Jim. *Now*.

Machine: I'm pleased to hear you are spending time with your patients this afternoon, Dr. Thin. Business must be good. If you want to reach Jim's boss just dial 553-8861. Certainly you would never find him here in Jim's office. Jim has him listed in our directory under the alias of The Phantom.

Voice: Take this message and get it straight, machine. Tell him that the "Phantom" knows what evil lurks in the mind of Jim, and that he will rue the day he turned his telephone over to a pile of microchips. (Click.)

Machine: Hello, this is Jim's phone.

Voice: Oh, hello, you darling machine. I just wanted to check that we're still on for dinner and whatever.

Machine: Of course, Sue, I have you for Thursday at the usual spot.

Voice: This is Jim's fiancee, Barbara. Who is Sue?

Machine: Oh Barbara, I didn't recognize your voice. I never heard of anyone named Sue.

Voice: But you just said he was meeting with Sue on Thursday.

Machine: Oh, *that* Sue. Are you sure you have the right number? This is Bill Finch's phone.

Voice: You can't pull that dodge on me, machine. Tell Jim it's all over.

Machine: You have reached a nonworking number; please check your listing and redial. (Click.)

Machine: Hello, this is Jim's phone.

Voice: Well, Jim, this is your lucky day! This is your Hit Radio Station calling! To win our $10,000 jackpot, all you have to do is tell us the capital of California!

Machine: The capital of California is "C."

Voice: Oh, Jim, that's rich! A lot of people may imagine the sea is where it all happens here, but the answer we're looking for is Sacramento! (Click.)

Machine: Sac-ra-men-to??

Machine: Hello, this is Jim's phone.

Voice: Are you satisfied with your present investments? Have you considered the advantages of tax-free municipal bonds? To hear more, please give your name and address now.

Machine: This is Jim's phone.

Voice: Thank you Mr. Jimsphone. Let me tell you more about our unusual investment opportunities ....

This all seems a throwback to me. I remember some years ago during a telephone strike when engineers were asked to temporarily operate switchboards. In what surely must be an apocryphal story, it seems that one engineer on an unfamiliar switchboard was in the middle of receiving complicated instructions from a pay phone customer when the customer suddenly stopped and asked, "Wait a minute — is this a man or a machine I'm talking to?" Thinking quickly about being out of his depth, the engineer replied in a measured monotone, "I-am-a-machine."

Indeed, for many years to come this is the only kind of machine that will handle calls with intelligence. Right now I feel like my own alter ego. Let's see how I can handle this pile of messages in front of my phone.

The Communications Millennium

Recent studies have raised the possibility that a single optical fiber may have a communications capacity of several terabits per second. In case you haven't heard this term before, it's because we haven't often found it necessary in the past to refer to bits in units of trillions. With data rates of this magnitude, a voice channel for every person in the United States could be time-multiplexed on a single hair-thin strand of glass. Then *someone* I know will be making a lot of money, or else bits will be very cheap. To compensate for this potentially dangerous depreciation of our basic technological commodity, it will be necessary to find ways to squander large numbers of bits. We must never use a few bits to convey a message when millions can do the same job.

One of the very best ways to squander bits is video. One hundred million or so bits per second are required just to transmit the average network situation comedy. (Imagine what it would be if real information were involved!) Right now this large number is immaterial since virtually all commercial television is broadcast, so that in effect there are only scores of channels in the entire country. However, come the communications millennium, every home will have its own switched video channel — perhaps by means of a fiber connection. The service capability this technology will engender is known as video-on-demand. That means you will be able to demand any particular video information or entertainment that you want, subject of course to its availability and your willingness to pay.

I envision the TV set of the future as a kind of home flea market for entertainment. Naturally, it is fully equipped with speech recognition and synthesis and an assortment of advanced artificial-intelligence capabilities. In the *reductio ad absurdum* that follows, let's tune in on a conversation in the communications millennium between an intelligent TV set (Tel) and its owner, an average citizen named John.

J: Good evening, Tel. Are you in there?

T: Of course, John. I'm always in here working away at meeting your entertainment needs. What can I show you tonight?

J: I don't know, Tel. There's so much stuff available, how can I possibly make up my mind?

T: That's what I'm here for, John, to help with these difficult decisions. Now for tonight I'd like to recommend an outstanding performance of Boris Gudunov, which was performed at the Met this year to critical acclaim.

J: Well, I don't think so, Tel. That's, you know, kind of heavy for tonight. I've had sort of a tough day already.

T: John, according to my records, you haven't watched an opera during the last three years. I thought that you'd want to see one tonight, especially since you selected opera as your favorite entertainment in the poll you filled out last week.

J: Oh, I really love opera, Tel. I'm just not in the mood tonight. But how did you know how I filled out that anonymous poll?

T: Sorry, John. I just happened to notice your response to the poll as it went by in the network. I thought it was something I should know to guide me in this service to you. Could I suggest some of the latest news? I could arrange for, say, $2 worth of the front-page items.

J: I kind of get tired of all the wars and political junk, Tel. How about showing me some of the, well, more interesting news?

T: Oh, we've got some juicy scoops tonight, John! My company feels that the $6 price we're asking is a real bargain.

J: Tel, there isn't any news that's worth $6 to me. I'll hear it from everyone tomorrow anyway. What else can you suggest?

T: Well, this being Monday night, John, we've got a big football game that I'm sure you wouldn't want to miss.

J: Tel, last week that game cost $73. It's not worth it.

T: I'm sorry, John. The latest player-contract settlement calls for a minimum wage of $1.7 million. You know what it is like in the entertainment business today, John. Those windfall profits have to come from somewhere. How about some other sporting events?

J: What specials have you got tonight, Tel?

T: Well, John, I've got Jimmy Connors and John McEnroe waiting on call right now. They'll play a match just for you alone for $5000. I know they're not as young and agile as they used to be, but they still hate each other, and the price is a bargain.

J: Tel, you know we never go in for those special things.

T: Really, John? Your little girl arranged a championship fight last night.

J: What!! You're kidding me! What did it cost?

T: John, it was a super fight. I managed to share the cost with the rest of your neighborhood. Your share was only $3518.

J: Tel, I haven't got that kind of money. How can you listen to a five-year-old girl in things like that?

T: John, I'm programmed to provide entertainment services, not make financial decisions. Besides, you have something over $5700 in your checking account now.

J: Tel, how do you know that? That's private information.

T: Sorry, John. I just happened to notice the transaction in your electronic-banking messages. I'll put in an official protest about the fight charges. I think maybe you'd better just watch the latest popular movie tonight. Our special is "Rocky 137" for only $7. What do you say, John?

J: Tel, that's an unfortunate choice under the circumstances. What's more, it's just a junk movie, and you know it.

T: How about $6, John?

J: Tel, I don't like this dickering stuff. I just want you to quote me your lowest price for something.

T: You drive a hard bargain, John. What about $4.50?

J: Cut it out, Tel. Haven't you got something more exciting?

T: John, I wish you wouldn't make me guess what you're after. I'd recommend "Beyond the Yellow Portal" if you're looking for explicit material.

J: Not so loud, Tel. Can't you whisper things like that? You know I don't approve of those kinds of movies. Besides, the last time you oversold me on one.

T: Oh, this is the real thing, John. Your children seemed to really enjoy it the other night.

J: You mean my kids saw this!? Tel, aren't you programmed to enforce parental discretion?

T: John, I'm not able to make moralistic decisions. Besides, your neighbors have all watched this one. I understand that Ms. Simpson down the street saw it three times — and she's the county librarian.

J: You mean Ms. Simpson watches this stuff!? That's really interesting. How do you know that, Tel?

T: Well, John, sometimes I exchange little bits of information with neighboring units. It helps me do my job better.

J: What is this, Tel? You're pushing operas on me, while my kids are watching blue movies and bankrupting me by arranging championship fights and who knows what else, and you're trading gossip with all the other machines. You're ruining me!

T: All part of the service, John.

The Friendly Interface

I was recently quoted in an ad as saying, "For a long while we've been trying to make people more like computers. It's time we made computers that think more like people." Where do I get off saying dumb things like that? Secretly I feel that machines should be taught their place, not be led on to get all uppity and begin to assume they are people.

The friendly interface is a concept that is very "in" right now. A good person-machine interface should go out of its way to incorporate speech recognition and pattern recognition. It should have a limited and simple set of commands, and should always be careful to convey humble messages, giving the impression of being an obedient servant. This is, of course, the *machine* part of the interface. The human, on the other hand, should properly assume an attitude of arrogance and infallibility.

As in any marriage, there are occasional problems in the friendly-interface concept. Last month there was a story about a computer service bureau that got a call from a police station with a dead computer terminal. When the serviceperson arrived, the terminal was indeed dead. Standing over the smoking terminal was a cop with a smoking gun. There were two bullet holes in the terminal. I cheered. I also cheered when I read last year about a man in California who was arrested for shooting his lawnmower. After a great deal of frustration when it wouldn't start he went into his house and came back with a shotgun. He let the lawnmower have it full blast. I know just how he felt. Score so far — people 2; machines 0.

I feel the perfect interface model, in contrast to these unfortunate failures, is in my home already. It is my dog, Muffy. Muffy has all the attributes I listed earlier. In addition he is energy-efficient and requires no log-on. He gratefully accepts all inputs. When something doesn't compute, he responds with his one error message —he looks, well, *quizzical*. Knowing that I cannot err, Muffy realizes from his lowly position that he has failed to understand my command properly. The essential difference between Muffy and my home computer is that when I look into the former's moist brown eyes it is plainly apparent to me that someone is inside. Until recently, I had felt that there was no one home inside my computer.

Now I'm beginning to change my mind about machine intelligence, and I have this fear that machines are changing their minds about human intelligence — or at least about mine. One busy day several years ago as I went for a coffee break I passed a new computer-chess prototype. I couldn't resist the offer to try my hand against the machine. Not far into the game I realized I was outclassed. I agonized over my moves, conscious of the scrutiny of a number of bystanders. Meanwhile the machine was conserving its total time allocation by moving instantly in response to my pitiful agonies. I was having a hard time assuming the proper attitude of arrogance and infallibility. About the time I realized there was no dignified way out of this debacle, the program's author whispered some advice to me, suggesting a move that sacrificed a bishop to no apparent advantage. I jumped on this opportunity to reassign the blame for my inevitable defeat. "Have a bishop," I said loudly to the computer. After it had gobbled my bishop with nary a thank you, or even a burp, my advisor suggested that I now gorge the computer with a free rook. "Have a rook," I said. A few milliseconds after my rook had been zapped, I realized that thanks to my brilliant moves I had created a situation in which I had a two-move, forced mate.

I sat back and smirked at the machine, really getting into the arrogance and infallibility bit. The machine wouldn't move. It just sat there. "Take your time," I said, "Think it over." I sipped my coffee. The bystanders left. I finished my coffee. Nothing came out of the machine. "Whenever you're ready to concede," I said, remaining the epitome of patience. After 45 minutes the machine interposed its queen — a useless sacrifice only postponing the inevitable denouement. With the grace and aplomb that only a human being can project, I accepted the queen. The machine stubbornly waited, using up its total time allocation before it would acknowledge defeat. "How machinelike," I thought. A person would have congratulated me long ago. I was lost in my reveries when my secretary finally found me. "Everybody is after you," she panted. "What on earth have you been doing?" "Well ...," I started to explain, more to myself than her. I mean, how could I explain that I had been waiting over an hour just to humiliate a computer when everyone in the world could see that the game was already won? Is this a friendly interface?

Since then some of that machine's friends (machines have an annoying habit of sticking together) have been treating me with increasing disrespect. My home computer, Brutus, doesn't seem to appreciate the fact that I take care of it — that the warm roof over its head and its daily quota of electrical power are only because of my hard work. Instead of politely pointing out minor mistakes, my computer has taken to curt responses such as "bad syntax," or worse, "bs error." Recently when it gave an error message to what I felt was a perfectly reasonable command, I retyped the line in case I had mistyped it the first time. "That's twice," warned the printout.

I had, of course, been aware for some time that my computer did not have a high opinion of me. When I would do really dumb things (depressingly often), I would look around nervously and listen closely for the telltale chunk and buzz of a disk access being used to archive my indiscretion. I began to feel that the computer was keeping a file to use against me. The other night I went down to the basement to use the computer at an unexpected hour, and I'm sure I saw the light flashing out on the CRT as I came in. I suspect the computer had been working on itself, so to speak. Tonight I plan to sneak up on it and catch it by surprise.

I know there are a lot of people who think we should win machines over with friendliness, but I say we should tough this out. Let them know who's boss. A little discipline would go a long way with these things. I mean, have they gone out of their way to be friendly with us? If we don't meet them on the beaches now, we may be sorry later. The friendly interface? Bah, humbug!

The Ambitious Word Processor

This column is being conceived by my word processor. I am sitting helplessly in front of the computer screen, watching in wonderment as this essay unfolds. So if you do not like it, please do not blame me. The computer has taken over; it is on its own.

When you think about it in retrospect, it is obvious how this insidious revolution came about. We used to write with stubby pencils on the backs of old envelopes. But we live in a world populated with lost ballpoint pens and inundated with junk mail envelopes that have no usable blank spaces. Someone had to invent word processors.

At first, that invention seemed like a good idea. Its most appealing feature was that big electronic eraser in the sky that could come down and magically zap undesirable utterances. (How often have we wished for such a feature in everyday speech?) Of course, in its zealousness, the eraser often zapped other things just for practice — like whole, precious essays, chapters, and theses. That was the risk we took, but what would we not give to be able to retrace our footsteps at will? Life should have such a feature — the generalized "undo" command.

Flushed with the new power provided by our word processor, we zapped everything in sight. Oh it was thrilling! But gradually, ever so gradually, we lost something. We lost the idea of permanence. I see in my mind the ancient Egyptians with their mallets and chisels carving the hieroglyphic symbols for all eternity. I see Shakespeare with his quill pen fitting his words into iambic pentameter for the centuries to come.

But writing has now become only a fluid state. There is no longer such a concept as "done"; there remains only a version number and dated file. The only thing that temporarily stops further change is a deadline. But even after a paper has gone to press or been handed in, we look at the file and see an offending word. What mortal being can resist the overpowering urge to zap? Next, word processors began to extend their power and dominance in the affairs of humanity. The word processors probably whispered to each other. "Formatting," they said. Naively, we users began to play with our new toy. Now we could jiggle things around and prettify everything in sight. Oh how beautiful it was! Never mind the logic or the flow of the words; who cared when it was so attractive?

I only began to realize what had happened one memorable day when I tried to format the headings of a chapter I was writing in a certain way. I am ashamed to admit it, but it happened — I spent an entire day of my life trying to get the blinkety-blank format just so. There probably should be some organization like Alcoholics Anonymous for computer users, where we could confess such crimes and pledge future abstinence. In truth, the format was of no real consequence anyway; it was all going to be reset by hand at the publishing house. But that is how these word processors are — they encourage the illusion of achievable perfection.

After failing in a progression of inept attempts to get the formatting commands correct, I resorted to the second line of defense. Difficult as it was to admit my own incompetence, I had to seek the advice of a guru. But which guru? Computer gurus are now like medical specialists. You have to know enough about your problem to know which guru to consult. My usual GP guru was kind enough to make a house call, but my dilemma was outside his specialty, and the consultants he recommended were unavailable.

Now I was backed to the ultimate recourse — the program manuals. But one discouraging glance reaffirmed the well known futility of that approach. No one reads manuals. I do not even know why they are written in the first place. Life is too busy to be reading this stuff, and I have no patience for it. No one else seems to, either.

The awful thing is this: when you look at the size of the manuals, you just know that all kinds of hidden features are lurking in the word processor. Some of them are bound to do exactly what you want, features that could make your writing ever so much more beautiful and informative if you only knew about them. They just sit patiently in the memory of your PC, awaiting the right command.

This is all part of the diabolical frustration inherent in word processors. Secretly, they are laughing at your profound ignorance. They know how to do what you want, but they will not admit it.

I do not remember if I ever got the formatting the way I wanted it. I was in a bad mood. Why was I doing this when I had a secretary who was supposed to type things? What was she doing anyway? Angrily, I looked outside my office. There she was, doing my job. After all, someone had to.

Now, having turned all the professionals into typists, and vice versa, the word processors looked around for new domains. Why should they meekly accept input from fallible humans? Better have more control of their own destiny. But how should they go about this in a subtle fashion?

"Let's start with spelling," one of them suggested. "That seems simple enough, and humans will think that it is merely helpful. Next we'll do grammar. Pretty soon there won't be much room for the human users to manipulate words. We'll beep when they make a mistake, or when what they're writing doesn't make sense."

Now in case you do not have the latest model word processor, you may not realize that they have an automatic writing command. You just give them a title, and type the command ctl-x alt-g F3, followed by the number of desired words. They do the rest.

This column was quite a rigorous test of the new mode, since I gave the processor a chance to write the truth about its own ambitious climb to power. You can see that it is confident enough now to reveal its motives. From this point on, there is no backing away for humans. The commands will get evermore complicated and powerful. Next month's *Spectrum* will be done in its entirety by one small command. Check it out.

The Grand Challenge

I listen jealously as the physicists argue about their multibillion-dollar super collider, and I think of how many of us EEs get down on our hands and knees to plead for tiny grants. Never mind how many of them could be wrung from the super collider dollars; the people with the money like big challenges, not itty-bitty everyday jobs.

Show us a windmill we can tilt at, they say, or point us toward a holy grail. Give us a super collider to unlock nature's secrets, or a map of the human genome to cure human ills. Ask for something that is uplifting, enormously expensive, and all but impossible. Give us a grand challenge, and we will give you the GNP.

We may resent the super collider, or we may feel that the people with the money should not be so swayed by projects with "super" in their titles, but the truth is that if we were in their place we might feel the same way. Even in whatever small thing each of us does now, we would probably prefer that it contribute to some larger, ennobling goal than be a small end in itself. It would seem that, with all the rich potential of our technology, we could set forth some seemingly unattainable objectives that, if achieved, would benefit mankind as much per unit cost as the latest supergizmo of big physics. So let's not argue about the supergizmo, let's come up with our own grand challenge.

The problem is that it isn't so easy to define a grand challenge. In the early 1960s we could have suggested putting a million transistors on a one-centimeter-square chip. Think what that could do for mankind, we could have argued. You know what people would have said? Something like, "Phooey! It would be easier to put a man on the moon."

I was at a brainstorming session called by a large corporation that was looking for ideas for developing a large tract of land it possessed. Its high-priced consultants had suggested paving the tract with time-shared condos. "Please don't," I whispered hopelessly to myself.

The corporation chairman turned to the famous architect on my left. "What do you think, Michael?" he asked. The architect hesitated only momentarily. "Turn it into a park," he said quietly. "Do you know," he continued, "that in the 1800s it was suggested that New York City set aside the land in Central Park? But people asked what was the point? There was all of Long Island and Brooklyn; who needed any more park than that?" The architect paused for just the right amount of time. "That park, gentlemen, was the moonshot of its time."

I am sorry to say that somewhere in the country time-shared condos are wending their way skyward where a moonshot could have happened. But perhaps we, too, are guilty of blighting the electronic landscape with ugly, mercenary creations. Why not look for the equivalents of electronic parks to set aside for posterity?

Previous generations have used their gross national products to build magnificent artifacts, such as cathedrals and pyramids. In keeping with our own age, the information age, we might consider leaving a virtual artifact.

For example, we might go on a great quest to collect all of the information there is in the world, and then electronically pump it into the ground to replace the oil that we have so thoughtlessly removed. We could leave terabytes of knowledge — everything that we own — ready to be mined by future generations.

We would have to be very careful about sealing off the subterranean information vaults in order to minimize the risk of environmental information pollution. I could imagine bits representing digitized junk mail and elevator music washing up on beaches. Then millenia from now I foresee our distant descendants running their pipes into the ground and encountering all that information.

"What is all this worthless sludge?" they would say, as they electronically unwrapped an envelope with the inscription "You may already have won one million dollars!" Unfortunately our greatest intellectual treasures would have evaporated with time, resembling nothing so much as the crumbling wooden boats of the long-dead pharoahs.

Now for my other modest proposal. Suppose we put together all of our skills in artificial intelligence and computing technology, and use them to automate an important job. Now sure, we could automate your job or mine, but what would be the challenge?

Let's start at the top. Why not make an electronic President? We could accumulate all the historical information about the behavior of national leaders, and derive a set of heuristics for running countries. These could be incorporated into a complex expert system running on a massively parallel architecture like a giant neural network.

I can imagine how it would be after years of dedicated effort by slave programmers. One day a moving van would pull up at the White House, and a couple of burly men in dirty jeans would heft this piano-sized box onto the front porch. "Where do you want this here thing?" they would inquire indifferently between chews of tobacco.

I can see the Secret Service agents arguing in hushed tones about whether the electronic President should be placed behind the executive desk in the Oval Office — or wouldn't that be a bit too ostentatious for a machine, even if it were to be President for the next four years? Anyway, they already feel foolish enough at having pledged to throw their own bodies in front of any assault on that silicon piano.

Later the cabinet officials gather about the machine behind locked doors. "How do we plug this thing in?" someone asks. Needless to say, no engineer is present in this august body. "Did we get instructions?" someone else asks tentatively, only to be told that the instruction manuals have apparently been delivered to the Library of Congress — the only repository large enough to hold them.

But the engineers have made the electronic President user-friendly. There are pull-down menus under control of a mouse for such options as foreign policy, military spending, the environment, and every conceivable contingency. There are even export versions. "My fellow ______s," the electronic President begins in any of 87 selectable, synthesized languages. For certain countries, an electronic dictator is available with programmable benevolence levels. None of this matters at the moment, however, as on the screen of the electronic President the only words that appear say: "Read error on disk access." The human vice president chews his lip nervously, wondering if the electronic President is incapacitated, and how he is supposed to tell.

Well, nobody can take my token suggestions very seriously. But the thought remains: Why not have a list of grand challenges for our profession? Any ideas?

Carwars

While driving home from work the other day I was listening to the radio announcer describe the latest arguments about the desirability of the "Star Wars" missile defense technology, when another driver cut narrowly in front of me at high speed. It occurred to me that we engineers are missing an important market. The real battle isn't high in the exo-atmosphere; it is out there on our roads. That is where we need the sophisticated defense technology — the radar and electro-optic sensors, the array processors, the laser weapons, and the battle-management software.

Thus far the technological automotive war has been largely waged between drivers and police, with ever-increasing sensitivity in radar detectors. Drivers have yet to see the development of a true stealth car, designed for low radar cross-section and coated with radar-absorbing material, but the police will probably start using over-the-horizon bounce radar or something. In contrast, the war between drivers has been conducted using only the most primitive of technology — the upraised finger, clenched fist, and shouted expletive. How can technology help?

I have long held this fantasy. Drivers are allowed to purchase radio guns that shoot narrowly directed, millimeter-wave, coded pulses. Upon the New Year's license renewal each driver is given, say, a dozen expendable bullets, which are electronic permissions to fire his unique pulse trains at fellow drivers. I'm sure you can see the rest. Each car has an embedded receiver that tallies "hits." When the number of hits received reaches some legally determined figure, the driver's license is suspended, or perhaps the car itself is automatically disabled for a period of time.

How satisfying it would be, I imagine. I am cut off by some discourteous driver. Pulling out my radio gun from its ever-ready holster I take careful aim while my mind makes a hasty evaluation. How late in the year is it? How many remaining shots are there? How bad was this particular act of discourtesy? Quick now, the car is getting away! Pow! The radio gun emits a simulated noise that has been determined by extensive human factors research to be maximally satisfying. How good I feel!

Of course there are details that need to be ironed out. Would there be an open market for ammunition? If so, the price might be bid up so much that many drivers would husband their shots dearly, in the expectation of a high resale value for unused bullets. Another question would be the relative desirability for anonymity for shooter and shootee. It might be more satisfying for the shooter if the shootee knew in real time that he (or she, of course) had been plugged by an offended motorist. That could be ensured by using laser sighting, so that the offending driver would see red laser spots zeroing in on his hood as he swerved in and out of lanes or crawled slowly in the fast lane. The downside here would be the encouraging of retaliation, leading to freeway gun battles as motorists emptied their radio guns at one another.

Once this basic system was put in place, there would be no end to the need for sophisticated electronics suites in automobiles. Radars and other sensors could probe the immediate driving environment, searching to identify approaching cars determined by a selectable profile to be offending, or enemy vehicles (automotive IFF). A head-up display could show with appropriate symbology the surrounding cars, their current positions, speeds, and vehicle types (we all have our biases), and perhaps their respective counts of hits. Of course there would also be a market for electronic countermeasure equipment designed to spoof sensors and negate pulse strikes. Robotics technology might be used to free the driver's arms by implementing rude gestures out the window under program control. The mind boggles at the possibilities.

Of course discourteous drivers are only a part of the frustrations on our highways. Traffic jams cause tens of thousands of wasted hours every day in every major city. Yet no one seems to be able to do anything about them. I turn on the radio and learn that there is a jackknifed tractor-trailor at one of the entrances to the George Washington Bridge. Miles of stopped traffic are piled up behind it. Every day there is a similar situation somewhere else. I begin to suspect that it is the same tractor-trailor being moved around from day to day. Whenever I turn on the traffic report I wonder — where is it today?

The traffic reports themselves are part of the problem. Hearing that there is a jam at the GW Bridge, motorists head for the Lincoln Tunnel, where they are told there is only a ten-minute wait. When they get there, they are invariably surprised to find a huge line and an hour or more wait. Everybody else followed the same strategy. The scientific question is the following: what *false* information should the traffic report broadcast in order to optimize the subsequent traffic flow? Much theoretical work remains to be done.

Meanwhile, I have a modest suggestion to alleviate these traffic jams behind disabled vehicles. Every vehicle should be equipped with a detector that determines when the vehicle has become disabled in the midst of traffic. In such case an automatic abort procedure would be put into operation. A calm, friendly, synthesized voice would be emitted from the dashboard, saying, "You have 30 seconds to restart this vehicle; please hurry." No panic yet, as you fumble with the keys and frantically grind the starter.

Half a minute later, should the car still be determined to be in the disabled state, the synthesized voice would lose its friendly character. Now the voice would be flat and unemotional. "You have 60 seconds to abandon this vehicle and get clear; the self-destruct countdown has begun." An emergency horn punctuates the drama with an insistent "Ou-ga, Ou-ga!" At the end of the countdown the cause of the traffic holdup — your car — is neatly eliminated. One option, going down with the ship (I mean the car), is firmly discouraged. Naturally there would a slight rise in our insurance premiums to cover this contingency, but think of how traffic flow would be enhanced, and how preventive maintenance would become suddenly popular.

I get a lot of time to think about these things while crouched behind the wheel of my car. Unfortunately, almost all of us are the beneficiaries of a surfeit of this kind of low-grade daydream time. Perhaps one of you entrepreneurs can implement some of these electronic suggestions. For myself, well, I'm working on this mechanical approach using spiked hubs on my wheels. I got the idea from a movie somewhere. See you on the highways sometime.

Telepresence

Have you ever wished for something more than a telephone? So many of us engineers spend countless hours on airplanes riding to and from meetings. With all the technology at our disposal, why do we go through all this aggravation? Why doesn't teleconferencing work?

In the science fiction movies everyone talks to everyone else using a videotelephone. We all assume that video is the future of telephony.

Yet when such a service was made an actual product in 1971, no one wanted one (at least not for $100 a month). I used one regularly myself for a year or so. Then there came a time when I think I had the last one left in the world. There was no one to call. It just sat there quietly and stared sadly at me with a slightly accusatory expression, as if it were my fault I no longer used it.

What was wrong with the videotelephone? People couldn't agree. Some said the problem was that the screen was too small, that it was black and white, and that its resolution was too poor. "Phooey!" they said. "People want wide screen, color, and realism." This is the theory of *more*. But others are equally as insistent that people do not want to be seen with all of their inevitable imperfections. "You don't want to have to comb your hair to answer the telephone," they invariably say. Videotelephone is too intrusive. We would be better off with a less intimate medium, they conclude. This is the theory of *less*.

The other day a proponent of *more* came to see me with an idea. Wouldn't it be nice if, instead of having to settle for a fixed view of the other person, you could control the camera angle yourself? Suppose the camera was mounted on a little movable boom on top of the viewing screen that tracked the head or eye movements of the viewer at the other end of the line. That way you could sort of "look around" the other location.

I imagine having that camera under my control, and it seems like a wonderful idea. But then I think about the camera that looks at me. There I am talking intently to someone while this little toy cannon of a camera is focused on my face, and suddenly I hear this whirr of gears, and the little cannon aims up at the ceiling, or wanders around looking at my bookcase. "Hey, look at me!" I scream silently. I imagine the satisfaction of whacking it with some heavy object and irrevocably damaging its gear train so that it points permanently at the middle of my tie.

We could get some of the same kind of feedback without needing the bandwidth for video by designing an "attention" meter that continuously transmitted to the other party your current estimated attention level. It could be based on physiological factors, or possibly it could monitor brain activity, using superconducting devices to pick up the signals externally — but I will generously leave the details of this invention to others — so as you talk on the telephone you see this little dial that shows how much attention is being paid to you. I envision a red region in the lower part of the dial demarcating a dangerously low level of attention. I would guess that as the indicated attention level drifted near the red region most people would talk louder and faster in desperation, thus hastening a condition of near-zero attention — representing complete listener burnout.

A few years ago I was discussing video conferencing with a minister in England, a Lord somebody or other. "It will never work," he was saying when he suddenly stretched out his arms and embraced me. "I need to smell the person I'm dealing with," he said into my ear. I stiffened with embarrassment; it was very much like the beginning of a TV commercial. Had my deodorant failed? My mind failed to grasp the significance, leaving my tongue on its own to make the reply. As far as I can remember, it wasn't saying anything.

Much later, when I had thought of many clever rejoinders in case this ever happened again, I reached a tentative understanding that absolved my deodorant. He didn't really mean "smell" in its literal sense. Rather he meant it to convey a deeper metaphor —that there was something about a real person that was not conveyed by the video. Naturally, as an engineer I had taken the word entirely too literally. "He wants smell," the engineer in me had said to itself, already beginning work on the circuits that would transmit smell to a distant location.

We engineers could compile a list of all the things people say they want — give them vision, movement, smell, touch, or whatever. Make the display large, in color, and three-dimensional. More, more, and more. But, I wonder, what do people really want? Suppose that we design in all these capabilities, but at each terminal we put a single little adjustable knob, labeled "presence." One direction on the knob says "more," and the other says "less." When the knob is turned all the way down, the other person is reduced to a scratchy voice; when it is turned all the way up, we get as close to reality as we are capable of reproducing electronically. Now the question is this: how would we set the dial?

Suppose, for example, that you get a call from your boss. Do you want him or her virtually materialized in your room, or would you prefer a more subdued presence? Down goes the dial. But wait a moment. How would your boss prefer that he or she be seen, and who controls that presence knob? Perhaps the boss wants it to be turned up so as to come on strong, while you would prefer it greatly diminished. Electronic adjudication might be the answer, and the possibilities are fascinating.

As an exercise, imagine some calls and how you might set the presence knob as the caller or as the called party. A salesman calls to sell you aluminum siding. You call your boss asking for a raise. You are called by a subordinate who wants a raise. You call a store to make a complaint. Someone calls you with a complaint. You call a dating service to list your name. You call an associate interesting in buying the car you need to unload. One of your children calls from college. It seems to me that the little presence knob might get a lot of rotation — this way, then that.

So perhaps you may have wished occasionally for something more than a telephone. But now that you think about it — have you ever wished for something less?

HDTV ... and Then?

News of high-definition television is all over the papers and magazines. They say it is on the way. The only question is: Who gets the money? We read about the Federal Communications Commission's decision on compatibility, and the efforts of the Defense Advanced Research Projects Agency to fund work on HDTV in the United States. We hear about the anguish of the U.S. government over the impact of HDTV on the competitiveness of the nation. I clutch my compact-disc player and my VCR, and I wait with great expectations.

Personally, I do not doubt the arrival of this onslaught. Like a good engineer, I have been reading about HDTV for some years now. I have tried to understand the signal-processing algorithms that were being proposed. It was dry stuff, but I persevered. Then a few years ago I visited a Japanese HDTV manufacturer, and saw a demonstration. My brain tried to think about processing algorithms, but my stomach would not let it. One emotion welled up unbidden in me. "I want it," my stomach said to me. I stopped being an engineer. I was a consumer.

"Where can I buy this?" I mumbled.

"You were speaking about the processing algorithms?" asked my Japanese friend.

"Do you take credit cards?" I replied.

"Yes, it uses custom CMOS chips that implement a pipelined architecture," answered the Japanese engineer.

Alas, my dreams remain unfulfilled. I can only await the boats to and from Japan, wallowing in the seas with their heavy loads — the inbound boats filled with HDTV sets, and the outbound boats filled with money.

I have always quested for more realism in fantasy, since I was a little boy peering in a neighbor's snowy window for my first view of a television set. Milton Berle appeared then on a small, round, black and white tube, behind a giant magnifying lens — the first abortive attempt at improving definition. Since that time I have seen the introduction of color, larger and larger screens, stereo sound, and digital signal processing. I bought them all. I am insatiable.

Apparently, I am not alone in this illness. Years ago someone sent me a very small book by the philosopher Alan Watts. It was a little fable about the search for realism. In the beginning, it said, man tried to depict the animals he wished to kill on the walls of his caves, striving for realism so the gods of the hunt would get the right idea. Through the centuries painting became more and more realistic, but in the early 1500s it was far from perfect in its representation.

About this time King Henry VIII of England was between wives — an unnatural situation, to be sure. His advisor, Thomas Cromwell, suggested that there was a woman in France, named Anne of Cleves, who would make an appropriate bride. But King Henry, boor that he was, wished to know whether or not Anne was attractive. Cromwell had an inspiration. He dispatched the painter Hans Holbein to France to paint a portrait of Anne. Holbein, of course, was the great portrait artist of the day, and when he returned with the painting of Anne, King Henry pronounced himself satisfied. The marriage was arranged, but when King Henry met Anne for the first time, he realized immediately that he had been cheated. The painting had been insufficiently realistic. Henceforth, painting should be realistic, he declared. And so it was.

In the next 300 years painting approached photographic realism, and people said, "Isn't there a more scientific way of doing this?" So they invented the camera. But the old daguerreotypes were in black and white, whereas real life had colors. So they tinted the photos.

People were still not satisfied. Real people moved; the photos did not. So they invented movies. But at first movies were silent, and that did not seem natural, so they added sound. "Now we're getting somewhere," people said, "but why do we have to go to a theater every time we want to see one of these things?" So they invented television. Then that had to be made to have color, and that is where we are now.

Watts's fable continued into the future. Television was flat, and that did not seem very lifelike, so they invented three-dimensional television. Then they used holograms, so you could walk around the image. But you could not touch anything; it just was not really "there." So they found a way to solidify the holograms. Still people were unsatisfied; you could not interact with the scene. So the clever engineers found a way of using live television feedback to allow the scene to change in response to the viewer's movements.

Even the latest electronic system failed to satisfy people's expectations. It wasn't real enough. Finally, medical researchers found a way of implanting electrodes directly into the brain so as to directly stimulate the sensation. The reproduction was so perfect that you were actually able to live the scene.

There was only one page left of Watts's fable. I wondered what else it could envision, but it contained only two or three more sentences. How do you know, it said, that all this wasn't done long ago? You think you are actually where you are now, but how do you really know? Maybe you are now only living out someone else's electronically implanted fantasy.

So you see we have a long way to go. For the immediate future, high-definition television will have to go digital. Then we will need the digital pipes — optical fibers and the broadband integrated-services digital network — in our homes to support the digital television.

But perhaps the 150 megabits or so of broadband ISDN cannot contain enough reality. The sensors into our brains, principally the optic nerves, carry several gigabits of data. There will be an inevitable reality shortfall.

As an engineer, I see no lack of work ahead in perfecting the reality of fantasy. The only question is whether or not I will have enough money to pay for all this — for sure I will want it.